Lecture Notes in Mathematics

A collection of informal reports and seminars
Edited by A. Dold, Heidelberg and B. Eckmann, Zürich

83

Oscar Zariski

Harvard University, Cambridge, Mass.

An Introduction to the Theory of Algebraic Surfaces

Notes by James Cohn, Harvard University, 1957 – 58

1969

Springer-Verlag Berlin · Heidelberg · New York

Library of Congress Catalog Card Number 68-59477 Printed in Germany. Title No. 3689

PREFACE

These are the lecture notes of a course which I gave at Harvard University in 1957-58. As the supply or the original mimeographed copies of these notes has been exhausted some years ago, and as there seemes to be some evidence of a continued demand, I readily accepted a proposal by the Springer Verlag to republish these old notes in the current series of "Lecture Notes in Mathematics." I refrained from making any changes or revisions, for I feel that these notes can best serve their purpose if they are published in their exact original form. The purpose of these notes is to acquaint the reader with some basic facts of the theory of algebraic varieties, and to do that by self-contained, direct and I would almost say - ad hoc methods of Commutative Algebra, without overwhelming the reader with a mass of material which has a degree of generality out of all proportion to the immediate object at hand. I should also mention, incidentally, that the title of these lecture notes is somewhat misleading, for only three of the sixteen sections (namely, sections 7, 14, and 19) deal specifically with algebraic surfaces; the remaining thirteen sections deal with varieties of any dimension.

Oscar Zariski
Harvard University
December, 1968

TABLE OF CONTENTS

§1. Homogeneous and non-homogeneous point coordinates.

Let k be our ground field; it need not be algebraically closed. A_n will denote an n-dimensional affine space, and S_n an n-dimensional projective space. These spaces have coordinates in a universal domain.

Def. 1.1: If $P = (x_1,\ldots,x_n) \in A_n$, then the local field, $k(P)$, of P/k is $k(x) = k(x_1,\ldots,x_n)$, and $\dim(P/k) = \text{t.d.}(k(x)/k)$.

This definition is independent of the choice of the affine coordinates in A_n/k.

Def. 1.2: If $P = (y_0,\ldots,y_n) \in S_n$, then the local field, $k(P)$, is the field generated over k by the ratios y_j/y_i where $y_i \neq 0$ and $\dim(P/k) = \text{t.d.}(k(P)/k)$.

This definition of $k(P)$ is independent of the choice of homogeneous coordinates in S_n/k.

Note: If we fix $y_i \neq 0$, then $k(P) = k(y_0/y_i,\ldots, y_n/y_i)$.

To see this, let $y_i = y_0$ and assume $y_1 \neq 0$. Then for any j,

$y_j/y_1 = (y_j/y_0)/(y_1/y_0)$, hence

$$y_j/y_1 \in k(y_1/y_0,\ldots, y_n/y_0).$$

Let H_{n-1} be a hyperplane in S_n, given by $\sum_{i=0}^{n} a_i Y_i = 0$, $a_i \in k$ (we then say H_{n-1} is rational over k), such that $P \notin H_{n-1}$. Then $S_n - H_{n-1} = A_n$ is an affine space and $P \in A_n$. Clearly then Def. 1.1 and Def. 1.2 lead to the same field $k(P)$.

For a given point P we may always assume, without loss of generality, that H_{n-1} is the hyperplane $Y_0 = 0$.

Def. 1.3: If $P \in A_n$, the ideal $I(P)$ in $k[X_1,\ldots, X_n]$ is the ideal $\mathfrak{p} = \{f(X) \mid f(x) = 0 \text{ where } P = (x)\}$, and $\dim \mathfrak{p} = \dim(P/k)$.

Def. 1.4: If $P \in S_n$, the ideal $I(P)$ in $k[Y_0,\ldots,Y_n]$ is the ideal $\mathfrak{p}^*$ generated by the forms $F(Y)$ such that $F(y) = 0$ where $P = (y)$; and $\dim \mathfrak{p}^* = 1 + \dim(P/k)$.

When $A_n = S_n - H_{n-1}$, we say A_n is an <u>affine representative</u> of S_n. Let A_n be an affine representative of S_n, i.e. $A_n = S_n - H_{n-1}$ where we can assume H_{n-1} is given by $Y_0 = 0$. For any given point P in A_n, we have two ideals:

$I(P) \subset k[X_1, \dots, X_n]$ (P regarded as a point of A_n)

$I^*(P) \subset k[Y_0, \dots, Y_n]$ (P regarded as a point of S_n).

Clearly $I(P) = \{F(1, X_1, \dots, X_n) \mid F(Y_0, \dots, Y_n) \in I^*(P)\}$.

<u>Def. 1.5</u>: The homogeneous coordinates y_i of $P \in S_n$ are <u>strictly homogeneous</u> coordinates of P/k if $t.d.(k(y_i)/k) = 1 + \dim (P/k)$.

<u>Prop. 1.6</u>: If $(y_0, \dots, y_n)$ are homogeneous coordinates of $P \in S_n$, the following statements are equivalent.

(a) The coordinates are strictly homogeneous.

(b) If for a given i, $0 \leqq i \leqq n$, $y_i \neq 0$, then y_i is transcendental over $k(P)$.

(c) If P^* denotes the point $(y_0, \dots, y_n)$ in A_{n+1}, then $I(P^*)$ is homogeneous.

Proof: It is clear that (a) $\Longleftrightarrow$ (b), for $k(y_0, \dots, y_n) = k(P)(y_i)$.

Let $\omega \in k(P)$. Then we can write $\omega = f(y)/g(y)$ where f and g are forms of the same degree. Hence if we have $\omega_0, \dots, \omega_m \in k(P)$, we can write $\omega_j = f_j(y)/f(y)$ where $f_0, \dots, f_m, f$ are forms of the same degree.

Assume $I(P^*)$ is homogeneous, and $y_i \neq 0$ for some i, $0 \leqq i \leqq n$. Assume $\omega_0 y_i^m + \omega_1 y_i^{m-1} + \dots + \omega_m = 0$ where $\omega_j \in k(P)$. By the preceding paragraph, we can write $\omega_j = f_j(y)/f(y)$ where the f_j and f are forms of degree h. Hence $f_0(Y)Y_i^m + f_1(Y)Y_i^{m-1} + \dots + f_m(Y) \in I(P^*)$ and each term has a different degree. Therefore $f_j(Y)Y_i^{m-j} \in I(P^*)$ for $j = 0, \dots, m$ and so $f_j(y) = 0$ for $j = 0, \dots, m$. Hence $\omega_j = 0$ for $j = 0, \dots, m$, and y_i is transcendental over $k(P)$.

Now assume (b) and let y_ν be transcendental over $k(P)$. Let $f(Y) = \sum_{i=\mu}^{m+\mu} f_i(Y)$ be a polynomial in $k[Y]$, where $f_i(Y)$ is a form of degree i, and assume $f(y) = 0$. Let $\omega_{m-i+\mu} = f_i(y)/y_\nu^i$, then $\omega_0, \ldots, \omega_m \in k(P)$, and $f(y) = 0$ means $\omega_0 y_\nu^{m+\mu} + \omega_1 y_\nu^{m+\mu-1} + \ldots + \omega_m y_\nu^\mu = 0$. Thus $\omega_j = 0$ for $j = 0,\ldots,m$; and so $f_i(y) = 0$ for all i.

2. Coordinate rings of irreducible varieties.

Let V be a variety defined and irreducible over k.

Def. 2.1: Let $V \subset A_n$, let $\mathfrak{p} = I(V)$ be the prime ideal of V in $k[X]$. Then the coordinate ring, $k[V]$, of V/k is the residue class ring $k[X]/\mathfrak{p}$, and the function field, $k(V)$, of V/k is the quotient field of $k[V]$.

If $x_i = \mathfrak{p}$-residue of X_i, then $k[V] = k[x]$ and $k(V) = k(x)$. Let $P = (x_1, \ldots, x_n)$. Clearly $P \in V$. We call P the canonical general point of V/k. Hence $k(V) = k(P)$. We set $\dim(V/k) = \dim(P/k)$.

Def. 2.2: If $V \subset S_n$ and $\mathfrak{p} = I(V)$ is the homogeneous prime ideal of V in $k[Y]$, then the homogeneous coordinate ring is $k[Y]/\mathfrak{p}$. If $y_i = \mathfrak{p}$-residue of Y_i, then $P = (y_0, \ldots, y_n)$ is the canonical general point of V/k. The function field, $k(V)$, of V/k is $k(P)$. $\dim(V/k) = \dim(P/k)$.

Note: The y_i are strictly homogeneous coordinates of P/k.

(This follows from Prop. 1.6b.)

$t.d.(k(y)/k) = 1 + \dim(V/k)$.

Prop. 2.3: $V \subset S_n$, let H_{n-1} be defined by $Y_o = 0$ and let $A_n = S_n - H_{n-1}$. Let $V_a = V - (V \cap H_{n-1})$. Then V_a is defined and irreducible over k, and $k(V) = k(V_a)$. The coordinates $x_1, \ldots, x_n$ of the canonical general point of V_a/k can be identified with the quotients $\frac{y_1}{y_o}, \ldots, \frac{y_n}{y_o}$, where $y_o, \ldots, y_n$ are homogeneous coordinates of the canonical general point of V/k.

3. Normal varieties.

Def. 3.1: Let $V \subset A_n$. Let $(x_1, \ldots, x_n)$ be the canonical general point of V/k and let $Q \in V$. Then the local ring, $\mathcal{O}_Q(V/k)$, of Q on V is the ring $\{f(x)/g(x) \mid f(X), g(X) \in k[X], g(Q) \neq 0\}$ In other words, $\mathcal{O}_Q(V/k)$ is the ring of all (rational) functions on V/k which are regular at Q.

Clearly $k[V] \subseteq \mathcal{O}_Q(V/k)$.

Def. 3.2: Let $V \subset S_n$, $(y) = (y_o, \ldots, y_n)$ the canonical general point of V/k and $Q \in V$. The local ring, $\mathcal{O}_Q(V/k)$, of Q on V/k is the ring $\{F(y)/G(y) \mid F, G$ forms of same degree in $k[Y], G(Q) \neq 0.\}$

Prop. 3.3: (a) Let $V \subset S_n$, let V_a be an affine representative of V and let $Q \in V_a$. Then $\mathcal{O}_Q(V/k) = \mathcal{O}_Q(V_a/k)$.

(b) $\mathcal{O}_Q(V/k) \subset k(V)$, and $k(V)$ is the quotient field of $\mathcal{O}_Q(V/k)$.

(c) If $V \subset A_n$, $Q \in V$ and $\mathfrak{p}$ is the prime ideal of Q in $R = k[V]$, then $\mathcal{O}_Q(V/k) = R_{\mathfrak{p}}$. The ideal of non-units in $\mathcal{O}_Q(V/k)$ is the set $\mathfrak{p}R_{\mathfrak{p}}$ of all rational functions on V/k which are regular at Q and vanish there.

Proof: (a) follows from Prop. 2.3. (b) and (c) are obvious.

Remark 3.4: k-isomorphic points of V/k have the same local ring.

Def. 3.5: If W is a subvariety of V, defined and irreducible over k, then the local ring, $\mathcal{O}_W(V/k)$, of W on V/k is the local ring $\mathcal{O}_Q(V/k)$ where Q is any general point of W/k.

Def. 3.6: V/k is normal at Q, $Q \in V$, if $\mathcal{O}_Q(V/k)$ is integrally closed. V/k is a normal variety if it is normal at each point.

Prop. 3.7: If $V \subset A_n$, then V is normal $\rightleftarrows$ $k[V]$ is integrally closed.

Proof: Since V is affine, it is known that $k[V] = \bigcap_{Q \in V} \mathcal{O}_Q(V/k)$.

Prop. 3.8: For a projective variety V to be normal it is necessary that each affine representative of V be normal, and it is sufficient that V admits a covering by affine normal varieties.

Proof: This follows from the fact that if V_a is an affine representative of V, then $\mathcal{O}_Q(V/k) = \mathcal{O}_Q(V_a/k)$, see Prop. 3.3a.

Def. 3.9: If $V \subset S_n$, then V is arithmetically normal if $k[V]$ is integrally closed.

Prop. 3.10: Arithmetic normality implies normality.

Proof: Let $V_a = V - (V \cap H)$ be an affine representative of V where H is given by $Y_o = 0$. Then $k[V_a] = k[y_1/y_o, \ldots, y_n/y_o]$. Let $\xi \in k(V) = k(V_a)$ where $\xi \in \overline{k[V_a]}$, the integral closure of $k[V_a]$. For a suitable integer s, $y_o^s \xi$ is an integral function of the y_i. Hence $y_o^s \xi = F(y) \in k[y]$.

Now $\xi = f(y)/g(y)$ where f and g are forms of degree t. Since $y_o^s F(y) - g(y)F(y) = 0$, each homogeneous part is zero. Thus we can assume $F(y)$ is a form of degree s. Therefore we can divide by y_o^s and so $\xi \in k[V_a]$.

4. Divisorial cycles on a normal projective variety V/k $(\dim(V) = r \geq 1)$

Def. 4.1: A prime divisorial cycle on V/k is a subvariety W of V such that W is defined and irreducible over k and $\dim(W) = r - 1$.

Def. 4.2: If K is a finitely generated extension field of k and $t.d.(K/k) = r$, then a prime divisor of K/k is a valuation v of K/k such that the residue field of v has transcendence degree $r - 1$ over k.

Prop. 4.3: If Γ is a prime divisorial cycle on V, then $\mathcal{O}_{\Gamma}(V/k)$ is the valuation ring of a prime divisor v_{Γ} of $k(V)/k$. Furthermore, v_{Γ} is a discrete, rank 1 valuation; and the residue field of v_{Γ} is $k(\Gamma)$.

Proof: Let H be defined by $Y_0 = 0$. Since we can assume $\Gamma \not\subset H$, let $V_a = V - (V \cap H)$ and $\Gamma_a = \Gamma - (\Gamma \cap H)$. Then $\dim(V_a) = r$ and $\dim(\Gamma_a) = r - 1$. Let $R = k[V_a]$; then R is integrally closed and noetherian. If $\mathfrak{p}$ is the prime ideal of Γ_a in R, then $\mathfrak{p}$ is a minimal prime ideal in R since $\dim(\Gamma_a) = r - 1$. We have $\mathcal{O}_{\Gamma}(V/k) = \mathcal{O}_{\Gamma_a}(V_a)/k) = R_{\mathfrak{p}}$, where $R_{\mathfrak{p}}$ is integrally closed and noetherian because R is. Furthermore, since $\mathfrak{p}$ is minimal in R, $\mathfrak{p}R_{\mathfrak{p}}$ is the only proper prime ideal in $R_{\mathfrak{p}}$. We have thus shown that $R_{\mathfrak{p}}$ is a Dedekind domain with only one proper prime ideal. Hence $R_{\mathfrak{p}}$ is a discrete valuation ring of rank 1.

Since $R/\mathfrak{p} = k[\Gamma_a]$ and $R_{\mathfrak{p}}/\mathfrak{p}R_{\mathfrak{p}}$ is the quotient field of $R/\mathfrak{p}$, we have the last statement.

The following is an easy consequence of Prop. 4.3 (but we shall not prove it here):

Cor. 4.4: With the assumptions and notations of Def. 4.2, every prime divisor of K/k is a discrete, rank 1 valuation, and its residue field is also finitely generated over k.

Def. 4.5: If $\xi \in k(V)$, $\xi \neq 0$, the integer $v_\Gamma(\xi)$ is the order of ξ on Γ. Γ is a prime null cycle of ξ if $v_\Gamma(\xi) > 0$, a prime polar cycle if $v_\Gamma(\xi) < 0$. If $\xi = 0$, we define $v_\Gamma(0) = +\infty$.

Prop. 4.6: Let $Q \in V \subset S^n$, and let $\xi \in k(V)$. If no prime polar cycle of ξ passes through Q, then ξ is regular at Q.

Proof: Let $\mathfrak{o} = \mathfrak{o}_Q(V/k)$. A prime cycle Γ corresponds to a minimal prime ideal $\mathfrak{p}$ of $\mathfrak{o}$. By assumption $\xi \in \mathfrak{o}_\mathfrak{p}$, hence $\xi \in \bigcap \mathfrak{o}_\mathfrak{p}$ where $\mathfrak{p}$ runs through the minimal prime ideals of $\mathfrak{o}$. Since V is normal, $\mathfrak{o}$ is integrally closed; and so $\bigcap \mathfrak{o}_\mathfrak{p} = \mathfrak{o}$. Hence $\xi \in \mathfrak{o}$.

Prop. 4.7: If ξ has no polar cycles, then ξ is a constant, i.e. ξ is algebraic over k.

Proof 1: If ξ has no polar cycles, then $\xi \in \bigcap_{Q \in V} \mathfrak{o}_Q(V/k)$; and if v is any valuation of $k(V)/k$, then $v(\xi) \geqq 0$. Assume ξ is transcendental over k, and consider η where $\eta = 1/\xi$. By the extension theorem there exists a valuation v such that $R_v \supset k[\eta]$ and M_v contains the principal ideal (η). Since $v(\eta) > 0$, $v(\xi) < 0$ which is a contradiction. Hence ξ is algebraic over k.

Proof 2: We have $\bigcap_{Q \in V_a} \mathfrak{o}_Q(V_a/k) = k[V_a]$. Let H_i be the hyperplane given by $Y_i = 0$, and let $V_a^{(i)} = V - (V \cap H_i)$ where $i = 0, \dots, n$. $k[V] = k[y_0, \dots, y_n]$ and $k[V_a^{(o)}] = k[x_1, \dots, x_n]$ where $x_i = y_i/y_o$.

By Prop. 4.6 and Prop. 3.3, we have $\xi \in k[x_1, \ldots, x_n]$, hence $\xi = F_0(y)/y_0^s$ where F_0 is a form of degree s. Similarly, for each i we can write $\xi = F_i(y)/y_i^s$. Hence ξy_i^s is a form of degree s in the y's. Let $m = (s-1)(n+1)+1$, and let $\omega_1, \ldots, \omega_N$ be the monomials in the y's of degree m. Then $\xi \omega_\nu$ can be written as a form of degree m in the y's: $\xi \omega_\nu = \Sigma c_{\nu\mu} \omega_\mu$, $c_{\nu\mu} \in k$. This gives an equation for ξ over k, namely $|c_{\nu\mu} - \delta_{\nu\mu}\xi| = 0$.

<u>Prop. 4.8</u>: If $\xi \in k(V)$, $\xi \neq 0$, then ξ has only a finite number of prime null cycles (or of prime polar cycles).

Proof: Use the fact that any hypersurface, not containing V, cuts V in a variety which has dimension $r - 1$, and which therefore has only a finite number of irreducible $(r-1)$-dimensional components.

<u>Def. 4.9</u>: A <u>divisorial cycle on V/k</u> is an element of the free (additive) group generated by the prime divisorial cycles. If $Z = \Sigma m_i \Gamma_i$, the sum being finite, and if $m_i \neq 0$, Γ_i is a <u>prime component</u> of Z. Notation: $[Z] = \bigcup_{m_i \neq 0} \Gamma_i$. If all $m_i \geqq 0$ and some $m_i > 0$, write $Z > 0$; Z is <u>effective</u> if $Z > 0$ or $Z = 0$.

<u>Def. 4.10</u>: If $\xi \in k(V)$, $\xi \neq 0$, then the <u>divisorial cycle of ξ</u> is $(\xi) = \Sigma_\Gamma v_\Gamma(\xi)\Gamma$. The <u>null divisor</u> of ξ = $\Sigma_{v_\Gamma(\xi)>0} v_\Gamma(\xi)\Gamma$. The <u>polar divisor</u> of ξ is $-\Sigma_{v_\Gamma(\xi)<0} v_\Gamma(\xi)\Gamma$.

<u>Note</u>: $(\xi) = 0 \rightleftarrows \xi$ is a constant.

If ξ is not a constant, then neither (ξ) nor $-(\xi)$ is effective. $(\xi\eta) = (\xi) + (\eta)$.

Def. 4.11: Two divisorial cycles Z_1 and Z_2 are linearly equivalent (notation: $Z_1 \equiv Z_2$), if $Z_1 - Z_2$ is the divisorial cycle of a function.

Prop. 4.12: Given Z, the set of all functions $\xi \in k(V)$ such that $(\xi) + Z \geqq 0$ is a vector space over k.

We denote this space by $\mathcal{L}(Z)$.

Proof: This follows from the following two facts: for $c \in k$, $(c\xi) = (\xi)$; and $v_\Gamma(\xi \pm \eta) \geqq \min\{v_\Gamma(\xi), v_\Gamma(\eta)\}$.

Prop. 4.13: If $Z_1 \equiv Z_2$, and if, say, $Z_1 - Z_2 = (\eta)$, then the map $\xi \to \xi\eta$, $\xi \in \mathcal{L}(Z_1)$, is a linear isomorphism of $\mathcal{L}(Z_1)$ onto $\mathcal{L}(Z_2)$.

Theorem 4.14: The space $\mathcal{L}(Z)$ is finite dimensional.

Proof (for the case $r = 2$ only): We can assume $\mathcal{L}(Z) \neq (0)$. Fix $\xi \in \mathcal{L}(Z)$ such that $\xi \neq 0$. Then we can write $(\xi) = X - Z$ where $X \geqq 0$, i.e. X is effective. Hence $X \equiv Z$ and so $\mathcal{L}(X) \approx \mathcal{L}(Z)$. Thus we can assume Z is effective.

Clearly $k \subset \bar{k} \subset k(V)$, hence $\bar{k}$ is finitely generated over k because $k(V)$ is. Since $\mathcal{L}(0) = \bar{k}$, we can therefore assume $Z \neq 0$. Then we can write

$$Z = \sum_{i=1}^{h} m_i \Gamma_i \ , \ m_i > 0 \ , \ h \geqq 1.$$

We shall prove the theorem by induction on $\sum m_i$.

We can assume that there exists f in $\mathcal{L}(Z)$ such that $v_{\Gamma_1}(f) = -m_1$, since otherwise $\mathcal{L}(Z) = \mathcal{L}(Z')$ where $Z' = (m_1-1)\Gamma_1 + m_2\Gamma_2 + \dots + m_h\Gamma_h$ and the result follows from the induction hypothesis.

Let $M = \{\xi/f \mid \xi \in \mathcal{L}(Z)\}$. Clearly $\dim(M) = \dim(\mathcal{L}(Z))$. Let us denote the elements of M by η. Since $v_{\Gamma_1}(\xi) \geq -m_1$, we have $v_{\Gamma_1}(\eta) \geq 0$ since $v_{\Gamma_1}(f) = -m_1$. Let $\mathrm{Tr}_{\Gamma_1}(\eta) = \bar{\eta} \in k(\Gamma_1)$ denote the restriction of η to Γ_1, i.e. the v_{Γ_1}-residue of η under the map $\phi : R_{\Gamma_1} \to R_{\Gamma_1}/M_{\Gamma_1}$ where R_{Γ_1} is the valuation ring

$\{z \in k(V) \mid v_{\Gamma_1}(z) \geq 0\}$ and $M_{\Gamma_1} = \{z \in k(V) \mid v_{\Gamma_1}(z) > 0\}$. Then $\phi : \eta \to \bar{\eta}$ is a mapping of M onto a subspace $\bar{M}$ of $k(\Gamma_1)/k$ where $\ker(\phi) = \{\xi/f \mid \xi \in \mathcal{L}(Z')\}$, $Z' = (m_1-1)\Gamma_1 + m_2\Gamma_2 + \dots + m_h\Gamma_h$. By our induction hypothesis, $\ker(\phi)$ is finite-dimensional. Therefore, it is sufficient to prove that $\bar{M}$ is finite-dimensional.

Let $(f) + Z = Y$. Clearly $Y \geq 0$. Since $v_{\Gamma_1}(f) = -m_1$, we see that Γ_1 is not a prime component of Y. Hence $\Gamma_1 \cap [Y] = \{P_i\}$, a finite set of (algebraic) points. We can construct two forms $F(y)$, $G(y) \in k[y] (= k[V])$ such that

(1) $F(P_i) \neq 0$ for all i; $F \neq 0$ on Γ_1; $F \neq 0$ on each prime component of Y

(2) $G = 0$ on each prime component of Y; $G \neq 0$ on Γ_1.

Since F^r and G^s (r, s positive integers) still satisfy (1) and (2), we can assume F and G have the same degree by choosing appropriate values for r and s. We have G/F in $k(V)$. Furthermore, each prime component of Y is a null cycle of G/F. Finally, G/F is regular at each P_i. These properties also hold for any power of G/F. Hence there exists an integer s which is large enough so that $\zeta = (G/F)^s$ has the same properties as G/F and the null cycle of ζ is $\geq Y$.

We can write $(\zeta) = Y + X - X'$ where $X, X' \geqq 0$. Let $\bar{\zeta} = \mathrm{Tr}_{\Gamma_1}(\zeta)$. Then $\bar{\zeta} \neq 0, \infty$ and $\zeta \in k(\Gamma_1)$.

Let $\bar{\mathfrak{p}}_1, \ldots, \bar{\mathfrak{p}}_q$ be the prime divisors of $k(\Gamma_1)/k$ which project into the set $\{P_i\}$. We shall show: if $\bar{\eta} \in \bar{M}$, then

(i) $v_{\bar{\mathfrak{p}}_i}(\bar{\eta}) \geqq - v_{\bar{\mathfrak{p}}_i}(\bar{\zeta})$

(ii) $v_{\bar{\mathfrak{p}}}(\bar{\eta}) \geqq 0$ if $\bar{\mathfrak{p}} \neq \bar{\mathfrak{p}}_i$, $i = 1, \ldots, q$

and this will prove the theorem (in view of known facts in the theory of algebraic functions of one variable).

The center of $\bar{\mathfrak{p}}_1$ is in the set $\{P_i\}$. From $(\eta) = Y' - Y$, $(\zeta) = Y + X - X'$, we see that $(\eta\zeta) = Y' + X - X'$. Since $P_1 \notin [X']$, $\eta\zeta$ is regular at P_1, $\overline{\eta\zeta}$ is also regular at P_1. Hence $v_{\bar{\mathfrak{p}}_1}(\overline{\eta\zeta}) \geqq 0$, and $v_{\bar{\mathfrak{p}}_1}(\bar{\eta}) \geqq -v_{\bar{\mathfrak{p}}_1}(\bar{\zeta})$.

Let Q be the center of $\bar{\mathfrak{p}}$ on Γ_1 where $\bar{\mathfrak{p}} \neq \bar{\mathfrak{p}}_i$, $i = 1, \ldots, q$. Then $Q \notin [Y]$ because $Q \notin \{P_i\}$ and $Q \in \Gamma_1$. Since $(\eta) = Y' - Y$, $Y' \geqq 0$, η is regular at Q. Hence $\bar{\eta}$ is also regular at Q, and $v_{\bar{\mathfrak{p}}}(\bar{\eta}) \geqq 0$.

5. Linear systems

Let $|Z|$ denote the set of all effective cycles linearly equivalent to Z. If $Z' \equiv Z$, then clearly $|Z'| = |Z|$. Let $\xi \in \mathcal{L}(Z)$, $\xi \neq 0$, then we can associate with ξ the effective cycle $(\xi) + Z \in |Z|$. It is clear that this map $\phi : \xi \to (\xi) + Z$ maps $\mathcal{L}(Z)$ onto $|Z|$. Assume k is maximally algebraic in $k(V)$. Then $\xi\phi = \xi_1\phi$ if, and only if, $\xi = c\xi_1$ where $c \in k$ (since $c = \frac{\xi}{\xi_1} \in k(V)$). Hence we can turn $|Z|$ into a projective space over k.

<u>Def. 5.1</u>: The set $|Z|$ together with the projective structure defined by $\phi : \xi \to (\xi) + Z$, $\xi \in \mathcal{L}(Z)$, $\xi \neq 0$, is called a <u>complete linear system</u> (the complete linear system determined by the cycle Z).

<u>Note</u>: The 1-dimensional subspaces of $|Z|$ are called <u>pencils</u>.

Viewing $|Z|$ as a projective space, we have $\dim |Z| = \dim \mathcal{L}(Z) - 1$, and $\dim |Z| = -1 \rightleftarrows |Z|$ is empty.

Let $Z' \equiv Z$ (whence $|Z'| = |Z|$), let $\phi' : \xi' \to (\xi') + Z'$, $\xi' \in \mathcal{L}(Z')$, $\xi' \neq 0$, and let $(\zeta) = Z - Z'$, $\zeta \neq 0$. Let $\psi : \xi \to \xi\zeta$, and define $\phi = \psi\phi'$. Then $\xi \in \mathcal{L}(Z)$ implies $\xi\zeta \in \mathcal{L}(Z')$. This shows that the projective structure is independent of the choice of Z in $|Z|$. Note also that $Z \not\equiv Z'$ implies $|Z| \cap |Z'| = \emptyset$.

<u>Def. 5.2</u>: A <u>linear system</u> is subspace of a complete linear system.

<u>Prop. 5.3</u>: If L is a non-empty linear system and X is any cycle which is linearly equivalent to the cycles of L, i.e. $L \subset |X|$, then the set $\mathcal{L}(L,X) = \{\xi \mid (\xi) + X \in L \text{ or } \xi = 0\}$ is a finite dimensional subspace of $k(V)/k$. Furthermore, $k \subset \mathcal{L}(L,X) \rightleftarrows X \in L$. Conversely, given any finite-dimensional subspace M of $k(V)/k$, and given any cycle X such that $(\xi) + X \geq 0$ for all $\xi \in M$, $\xi \neq 0$, then the set $\{(\xi) + X \mid \xi \in M, \xi \neq 0\}$ is a linear system L and $M = \mathcal{L}(L,X)$.

Proof: Since L is a subspace of X, $\mathcal{L}(L,X)$ is a subspace of $\mathcal{L}(X)$ which proves the first statement. The second assertion is obvious, and the last follows from the fact that M is a subspace of $\mathcal{L}(X)$.

We say $\mathcal{L}(L,X)$ is a <u>defining function module</u> of L.

<u>Notation</u>: Let $Z = \sum_{\Gamma} m_{\Gamma}(Z)\Gamma$; $m_{\Gamma}(L) = \min\{m_{\Gamma}(Z) \mid Z \in L\}$.

Let L be a linear system contained in $|X|$. Let Z_1 and Z_2 be linearly independent cycles in L; and let $Z \in L$ be linearly dependent on Z_1 and Z_2, i.e. Z is in the pencil determined by Z_1 and Z_2. Then, if $(\xi_1) + X = Z_1$, $(\xi_2) + X = Z_2$, $\xi_1, \xi_2 \in \mathcal{L}(X)$, we must have $(c_1 \xi_1 + c_2 \xi_2) + X = Z$ for some $c_1, c_2 \in k$. For any prime cycle Γ, we have $m_\Gamma(Z_1) = v_\Gamma(\xi_1) + m_\Gamma(X)$, $m_\Gamma(Z_2) = v_\Gamma(\xi_2) + m_\Gamma(X)$, and $m_\Gamma(Z) = v_\Gamma(c_1\xi_1 + c_2\xi_2) + m_\Gamma(X) \geq \min\{v_\Gamma(\xi_1), v_\Gamma(\xi_2)\} + m_\Gamma(X)$. Hence $m_\Gamma(Z) \geq \min\{m_\Gamma(Z_1), m_\Gamma(Z_2)\}$.

Assume $X \equiv X'$ while $X \neq X'$. Since $m_\Gamma(X) = m_\Gamma(L) - \min\{v_\Gamma(\xi) \mid \xi \in \mathcal{L}(L,X)\}$, we see that $\mathcal{L}(L,X) \neq \mathcal{L}(L,X')$ provided $L \neq \emptyset$.

<u>Prop. 5.4</u>: Let L be a linear system, and let Y_0 be a cycle such that $Z + Y_0 \geq 0$ for all Z in L. Then the set $L' = \{Z + Y_0 \mid Z \in L\}$ is a linear system and $\dim L = \dim L'$.

Proof: Fix X such that $L \subset |X|$. Let $M = \mathcal{L}(L,X)$. Then $L' = \{(\xi) + X + Y_0 \mid \xi \in M,\ \xi \neq 0\}$, and $M = \mathcal{L}(L', X + Y_0)$.

<u>Def. 5.5</u>: A positive cycle B is a <u>fixed component</u> of a linear system L if $Z \geq B$ for all $Z \in L$.

<u>Cor. 5.6</u>: If B is a fixed component of L, then the set $\{Z - B \mid Z \in L\}$ is a linear system having the same dimension as L.

Proof: By definition $Z - B \geq 0$ for all $Z \in L$. The result now follows from Prop. 5.4 with $-B = Y_0$.

Let M be a finite dimensional subspace of the vector space $k(V)/k$. Then there exists a smallest cycle X_0 such that $(\xi) + X_0 \geq 0$ for all $\xi \in M$. Let $L_0 = \{(\xi) + X_0 \mid \xi \in M,\ \xi \neq 0\}$. Then L_0 has no fixed component, and we have $M = \mathcal{L}(L_0, X_0)$. If L is any other linear system which admits M as defining function module, i.e., if $M = \mathcal{L}(L,X)$, then $X = X_0 + B$, $B > 0$, and B is a fixed component of L. Thus L_0

is the only linear system, <u>free from fixed components</u>, which admits M as defining function module. We will denote L_0 by $LS(M)$.

<u>Theorem 5.7</u>: (Theorem of the Residue (Restsatz)). Let L be a linear system and Y_0 an effective cycle. Let $L' = \{Z - Y_0 \mid Z \in L, Z \geq Y_0\}$. Then L' is a linear system. If L is complete, then $L' = |Z - Y_0|$.

Proof: Let $L_1 = \{Z \mid Z - Y_0 \geq 0, Z \in L\}$. Then $L_1 \subset L$ and, by Prop. 5.4, it is sufficient to prove that L_1 is a linear system. Let $Z_1, Z_2 \in L_1$, and let Z belong to the pencil determined by Z_1 and Z_2 in $|Z_1|$. We have $v_\Gamma(Z) \geq \min\{v_\Gamma(Z_1), v_\Gamma(Z_2)\}$, hence $Z - Y_0$ is effective since $Z_1 - Y_0$ and $Z_2 - Y_0$ are effective. Thus $Z \in L_1$, and this shows that L_1 is a linear system.

Assume L is complete, and let $X_0 \in L'$. Then $X_0 = Z - Y_0$ where $Z \in L$. Let $X \equiv X_0$, $X \geq 0$. Then $X + Y_0 \equiv X_0 + Y_0 = Z$ and so $X + Y_0 \in L$. Hence $X \in L'$, showing that $L' = |X_0|$.

Let L be a non-empty linear system, and Γ a prime cycle which is not a fixed component of L. Fix a cycle Z_0 linearly equivalent to the cycles in L and such that Γ is not a component of Z_0. (For instance, take Z_0 in L.) Let $M = \mathcal{L}(L, Z_0) = \{\xi \mid (\xi) + Z_0 \in L\}$. Let $\bar{\xi} = \mathrm{Tr}_\Gamma \xi$ for any $\xi \in M$ ($\bar{\xi} \neq \infty$), and let $\bar{M}$ be the set of all the $\bar{\xi}$. Since Γ is not a fixed component of L, there exists $X \in L$ such that $(\xi) = X - Z_0$ satisfies the condition $\mathrm{Tr}_\Gamma \xi \neq 0$. Hence $\bar{M} \neq (0)$. (Clearly $\bar{M} \subset l(\Gamma)$.)

$LS(\bar{M})$ is a linear system on Γ without fixed components. It is easy to see that $LS(\bar{M})$ depends only on L and Γ and not on Z_0. We shall refer to $LS(\bar{M})$ as the <u>reduced trace of</u> L <u>on</u> Γ (notation: $\mathrm{Red\,Tr}_\Gamma L$).

Theorem 5.8: Let $L' = \{ Z - \Gamma \mid Z \in L,\ Z \quad \Gamma \geqq 0 \}$ where L, Γ are as above. Let $\bar{L} = \text{Red Tr}_{\Gamma} L$. Then
$\dim L = \dim L' + \dim \bar{L} + 1$.

6. Divisors on an arbitrary variety V.

Let $V \subset S_n$, and let $(y_o, \ldots, y_n)$ be a general point of V/k. Let v be a valuation of $k(V)/k$. Then for some j, $v(y_i/y_j) \geq 0$ for all i. Let z_i be the v-residue of y_i/y_j, $i = 0, \ldots, n$; and let Δ be the residue field of v. Clearly the point $P = (z_o, \ldots, z_n)$ belongs to V. Let W/k be the locus of P over k, i.e. the irreducible variety with P as general point over k $(W \subset V)$. Then W depends only on v and V.

Def. 6.1: W is called the center of v on V.

Prop. 6.2: The following properties characterize the center W of v on V:

(a) $\mathcal{O}_W(V/k) \subset R_v$ where R_v is the valuation ring of v.

(b) $\mathfrak{m}_W(V/k) = M_v \cap \mathcal{O}_W(V/k)$.

Proof: Let W be the center of v on V. We know that $\mathcal{O}_W(V/k) = \mathcal{O}_P(V/k)$. Let $\mathcal{O} = \mathcal{O}_P(V/k)$, and let $F(y)/G(y) \in \mathcal{O}$, where F and G are forms of like degree s. We have $G(z) \neq 0$, and the v-residue of $F(y)/G(y)$ is equal to $F(z)/G(z) \neq 0$, showing that $F(y)/G(y) \in R_v$. Furthermore, this v-residue is zero if and only if $F(z) = 0$, i.e., if and only if $F(y)/G(y) \in \mathfrak{m}_W$. Hence conditions (a) and (b) are satisfied. Conversely, assume that conditions (a) and (b) are satisfied. Let $(z_o, z_1, \ldots, z_n)$ be a general point of W/k. We may assume, without loss of generality, that $y_o \neq 0$. Then $y_i/y_o \in \mathcal{O}_W(V/k) \subset R_v$, showing that if z'_i is the v-residue of y_i/y_o and W' is the center of v on V, then $(z'_o, z'_1, \ldots, z'_n)$ is a general point of W'/k. If $f(z) = 0$ is a homogeneous relation, of degree s, between the z's, with coefficients in k, then $f(y)/y_o^s \in \mathfrak{m}_W \subset M_v$, whence $f(z') = 0$. Conversely, if $f(z') = 0$, then $f(y)/y_o^s \in M_v \cap \mathcal{O}_W(V/k) = \mathfrak{m}_W$, whence $f(z) = 0$. This shows that $W = W'$.

<u>Def. 6.3</u>: Let V_a be an affine representative of V. Then v is <u>finite</u> on V_a if $k[V_a] \subset R_v$.

<u>Prop. 6.4</u>: If v is finite on V_a, and if W is the center of v on V, then $W \cap V_a$ is an affine representative W_a of W called the <u>center of</u> v <u>on</u> V_a. The prime ideal of W_a in $k[V_a]$ is $\mathfrak{m}_v \cap k[V_a]$.

Proof: The first statement is obvious, and the second follows from Prop. 3.3 and Prop. 6.2.

Since $k \subset \Delta$ and $\dim V = r$, we have $0 \leq \text{t.d. } \Delta/k \leq r - 1$; v is a prime divisor of $k(V)$ if $\text{t.d. } \Delta/k = r - 1$.

<u>Prop. 6.5</u>: If W is the center of v on V, then $k(W)$ can be identified with a subfield of Δ. Hence $\dim W \leq \dim v$.

Proof: This follows directly from Prop. 6.2.

<u>Def. 6.6</u>: v is of the <u>first kind</u> with respect to V if $\dim W = \dim v$.

If V/k is a normal variety, then we have a (1-1)-correspondence between the (r-1)-dimensional irreducible subvarieties Γ of V, i.e., the prime divisorial cycles, and the prime divisors v of the first kind with respect to V, where Γ corresponds to v if and only if Γ is the center of v on V. If V is not normal, we shall continue to call "prime divisorial cycle" of V any (r-1)-dimensional irreducible subvariety of V/k, but this time we may have several prime divisors of $k(V)/k$ which have a given prime divisorial cycle of V/k as center.

Theorem 6.7: (a) Let Γ be a prime divisorial cycle on V/k. Then every valuation of $k(V)$ with center Γ is a prime divisor of $k(V)/k$.

(b) There exists at least one and at most a finite number of prime divisors of $k(V)/k$ having center Γ on V/k. Furthermore, there is only a finite number of Γ's on V/k which are centers of more than one prime divisor.

(c) If Γ is the center of a prime divisor v, then $[\Delta : k(\Gamma)] < \infty$.

Proof: (a) By Prop. 6.5 we have $\dim \Gamma \leq \dim v$. Since $\dim \Gamma = r-1$, we have $\dim v = r-1$. Therefore v is a prime divisor.

(b) The existence of one prime divisor with center Γ is a well-known result. Let $\bar{\mathcal{O}}$ be the integral closure of $\mathcal{O}(= \mathcal{O}_\Gamma(V/k))$ in its quotient field. Let $\mathfrak{m}$ be the maximal ideal of $\mathcal{O}$. Since $\bar{\mathcal{O}}$ is a finite $\mathcal{O}$-module, the residue class ring $\bar{\mathcal{O}}/\mathfrak{m}\bar{\mathcal{O}}$ is a finite dimensional vector space over the field $\mathcal{O}/\mathfrak{m}$. Therefore the ideals of $\bar{\mathcal{O}}/\mathfrak{m}\bar{\mathcal{O}}$ satisfy the descending chain condition. Hence $\bar{\mathcal{O}}/\mathfrak{m}\bar{\mathcal{O}}$ has only a finite number of maximal ideals; and since each maximal ideal of $\bar{\mathcal{O}}$ contains $\mathfrak{m}\bar{\mathcal{O}}$, we have shown that $\bar{\mathcal{O}}$ has only a finite number of maximal ideals. If v is a prime divisor with center Γ and M_v is its maximal ideal, then $M_v \cap \bar{\mathcal{O}}$ is a maximal ideal of $\bar{\mathcal{O}}$, since its contraction in $\mathcal{O}$ is the maximal ideal of $\mathcal{O}$. If $\bar{\mathfrak{m}}$ is a maximal ideal of $\bar{\mathcal{O}}$, then the quotient ring $\bar{\mathcal{O}}_{\bar{\mathfrak{m}}}$ is a discrete valuation ring, and the corresponding valuation of $k(V)/k$ is a prime divisor, and is the only prime divisor v such that $M_v \cap \bar{\mathfrak{o}} = \bar{\mathfrak{m}}$. This proves that there are only a finite number of prime divisors with center Γ.

If H is a hyperplane in S_n, $V \not\subset H$, then $H \cap V$ contains only a finite number of prime divisorial cycles. It is sufficient, therefore, to prove the last part of (b) for prime divisorial cycles of the affine representative $V_a = V - (V \cap H)$ of V. Let $R = k[V_a]$, and let R' be the integral closure of R in its quotient field. Since R' is a finite R-module, we can write $R' = \Sigma_{i=1}^{t} R\omega_i$ where $\omega_i = f_i/g$, $i = 1, \ldots, t$, with $f_1, \ldots, f_t, g \in R$. Let $\mathfrak{p}_1, \ldots, \mathfrak{p}_s$ be the minimal prime ideals of R which contain g, and let $\Gamma_1, \ldots, \Gamma_s$ be the prime divisorial cycles determined by $\mathfrak{p}_1, \ldots, \mathfrak{p}_s$. If Γ is any prime divisorial cycle on V distinct from the Γ_i, then the minimal prime ideal $\mathfrak{p}$ of Γ in R does not contain g. Therefore $\mathfrak{o}_\Gamma(V/k) \supset R'$, and from this follows at once that if we set $\mathfrak{p}' = \mathfrak{m}_\Gamma \cap R'$, then $\mathfrak{o}_W(V/k) = R'_{\mathfrak{p}'}$. Hence $\mathfrak{o}_\Gamma(V/k)$ is integrally closed in its quotient field (which is the quotient field of R). Hence Γ is the center of exactly one prime divisor. This proves (b).

(c) is obvious.

We wish now to extend the results of §4 to arbitrary varieties V. However, we have already observed that a prime divisorial cycle may be the center of more than one prime divisor (of the first kind with respect to V); hence the free group generated by the prime divisors is no longer isomorphic to that generated by the prime divisorial cycles. Therefore, we must now replace the "prime divisorial cycles" of §4 by the "prime divisors" (of the first kind with respect to V), and the elements ("divisorial cycles") of the free group generated by the former, by the elements ("divisors") of the free group generated by the latter.

For clarity we restate Def. 4.5:

Def. 4.5': If $\xi \in k(V)$, $\xi \neq 0$, the integer $v(\xi)$ is the order of ξ on v; v is a prime null divisor of ξ if $v(\xi) > 0$, a prime polar divisor if $v(\xi) < 0$. If $\xi = 0$, we define $v(0) = +\infty$.

With the above indicated change, all the propositions of §4 remain valid, except Prop. 4.6, in which the conclusion "ξ is regular at Q" must be replaced by "ξ belongs to the integral closure of $\mathcal{O}_Q(V/k)$".

7. Intersection theory on algebraic surfaces (k algebraically closed)

Let $W \subset V$ where $\dim V = r$ and $\dim W = s$. It is well known that we can find maximal chains $W < W_1 < W_2 < \dots < V$ of irreducible varieties W_i where $\dim W_i = s+i$. Hence, if $\mathcal{O} = \mathcal{O}_W(V/k)$, then the Krull dimension of $\mathcal{O}$ is r-s.

Def. 7.1: W is a simple subvariety of V/k if $\mathcal{O}_W(V/k)$ is a regular local ring, i.e., if $\mathfrak{m} = \mathfrak{m}_W(V/k)$ has a basis of r-s elements. Equivalently

$$\dim_{\mathcal{O}/\mathfrak{m}} (\mathfrak{m}/\mathfrak{m}^2) = r-s.$$

Let $P \in V$ and let W be the variety for which P is general point. Then we say that P is simple if W is.

Cor. 7.2: If V is normal and $\dim W = r-1$, then W is a simple subvariety of V/k.

Since a regular local ring is integrally closed, we have

Cor. 7.3: Any non-singular variety is normal.

Def. 7.4: If $Z = \Sigma\, m_{\Gamma}(Z)\Gamma$ is a divisorial cycle on V, and if $P \in V$, then the local P-component Z_P of Z is $\Sigma_{P\in\Gamma}\, m_{\Gamma}(Z)\Gamma$. We say Z is locally linearly equivalent to zero at P if $Z_P = (\xi)_P$ for some function ξ.

If ξ exists, it is determined to within a factor which is a unit in $\mathcal{O}_P$. We say " $\xi = 0$ is a local equation of Z at P."

Let V be non-singular, let $P \in V$ and let Γ be a prime cycle through P. Then $I(\Gamma) = \mathfrak{p}$ is a minimal prime ideal in $\mathcal{O}_P$. Hence $\mathfrak{p} = \mathcal{O}_P t$, i.e., $\mathfrak{p}$ is principal, and $(t)_P = \Gamma$. (We recall that the local ring of a simple point is a unique factorization domain, hence every minimal prime ideal in that ring is principal.)

Prop. 7.5: On a non-singular variety V, every divisorial cycle Z is linearly equivalent to zero at every point P.

Proof: Since $Z_P = \Sigma_{P\in\Gamma}\; m_{\Gamma}(Z)\Gamma$, the result follows directly from the preceding paragraph.

Let F/k be a non-singular surface.

Def. 7.6: Let Γ, Δ be distinct irreducible curves on F/k. Let $P \in F$, and let $\xi = 0$, $\eta = 0$ be local equations of Γ, Δ respectively at P ($\xi, \eta \in \mathcal{O}_P$). Then we define

$$i(\Gamma,\Delta;P) = \dim_k(\mathcal{O}/\mathcal{O}(\xi,\eta)), \qquad (\mathcal{O} = \mathcal{O}_P),$$

and we call the integer $i(\Gamma,\Delta;P)$ the intersection multiplicity of Γ and Δ at the point P.

If P is not a common point of Γ and Δ, then either ξ or η is a unit in $\mathcal{O}_P$ and the intersection multiplicity is zero. If P is a common point of Γ and Δ, then P is a rational point, $\dim P/k = 0$, $\dim \mathcal{O} = 2$ and $\mathcal{O}(\xi,\eta)$ is an $\mathfrak{m}$-primary ideal ($\mathfrak{m}^{\rho} \subset \mathcal{O}(\xi,\eta)$

whence $0 < \dim_k \mathcal{O}/\mathcal{O}(\xi,\eta) < \infty$). We thus have

Cor. 7.7: $i(\Gamma,\Delta;P) \geq 1 \rightleftarrows P \in \Gamma \cap \Delta$.

Let $X = \sum_{i=1}^{r} a_i \Gamma_i$, $Y = \sum_{j=1}^{s} b_j \Delta_j$ be divisorial cycles without common components. Then we define $i(X,Y;P) = \sum_{i,j} a_i b_j i(\Gamma_i,\Delta_j;P)$. We define the intersection number $(X.Y) = \sum_{P \in F} i(X,Y;P)$, where the right-hand side is a finite sum since X and Y do not have any common components.

Theorem 7.8: Let Γ be a prime cycle, let $P \in \Gamma$ and let $\eta \in k(F)$, $\eta \neq 0$, such that Γ is not a prime component of (η). Let $\bar{\eta} = \mathrm{Tr}_{\Gamma}\, \eta$ ($\bar{\eta} \in k(\Gamma)$, $\bar{\eta} \neq 0$). Then $i((\eta),\Gamma;P) =$ degree of $(\bar{\eta})_P$. [Here, if $v_1, \ldots, v_g$ are the valuations of $k(\Gamma)/k$ which have center P on Γ and if $v_i(\bar{\eta}) = n_i$, then $\deg(\bar{\eta})_P = n_1 + n_2 + \ldots + n_g$.]

Proof: We can assume $\eta \in \mathcal{O}_P$. Let $\mathcal{O} = \mathcal{O}_P$, and let $\xi = 0$ be a local equation of Γ at P; then $\mathcal{O}\xi$ is the prime ideal of Γ in $\mathcal{O}$. We have $i((\eta),\Gamma;P) = \dim_k \mathcal{O}/\mathcal{O}(\xi,\eta) = \dim_k \bar{\mathcal{O}}/\bar{\mathcal{O}}\bar{\eta}$ where $\bar{\mathcal{O}} = \mathcal{O}/\mathcal{O}\xi = \mathcal{O}_P(\Gamma/k)$. Let $\bar{\mathcal{O}}'$ be the integral closure of $\bar{\mathcal{O}}$ in $k(\Gamma)$; $\bar{\mathcal{O}}'$ is a semi-local ring with a finite number of prime ideals $\mathfrak{p}_1, \ldots, \mathfrak{p}_g$ where $\mathcal{O}'_{\mathfrak{p}_i} = R_{v_i}$ for $i = 1, \ldots, g$. Each $\mathfrak{p}_i$ is maximal, and so $\bar{\mathcal{O}}'$ is a Dedekind domain. Therefore $\bar{\mathcal{O}}'\bar{\eta} = \mathfrak{p}_1^{n_1} \cdots \mathfrak{p}_g^{n_g}$. A well-known result allows us to conclude that $n_1 + \ldots + n_g = \dim_k \bar{\mathcal{O}}'/\bar{\mathcal{O}}'\bar{\eta} = \deg(\bar{\eta})_P$.

We shall show in a moment that $\dim_k \bar{\mathcal{O}}'/\bar{\mathcal{O}} < \infty$. Assuming this, we have

$$\dim_k \bar{\mathcal{O}}'/\bar{\mathcal{O}}\bar{\eta} = \dim_k \bar{\mathcal{O}}'/\bar{\mathcal{O}}'\bar{\eta} + \dim_k \bar{\mathcal{O}}'\bar{\eta}/\bar{\mathcal{O}}\bar{\eta} = \dim_k \bar{\mathcal{O}}'/\bar{\mathcal{O}}'\bar{\eta} + \dim_k \bar{\mathcal{O}}'/\bar{\mathcal{O}} .$$

Furthermore $\dim_k \bar{\mathcal{O}}'/\bar{\mathcal{O}}\bar{\eta} = \dim_k \bar{\mathcal{O}}'/\bar{\mathcal{O}} + \dim_k \bar{\mathcal{O}}/\bar{\mathcal{O}}\bar{\eta}$. Hence

$$\dim_k \bar{\mathcal{O}}/\bar{\mathcal{O}}\bar{\eta} = \dim_k \bar{\mathcal{O}}'/\bar{\mathcal{O}}'\bar{\eta} .$$

Thus $i((\eta),\Gamma;P) = \dim_k \bar{\mathfrak{o}}/\bar{\mathfrak{o}}\bar{\eta} = \dim_k \bar{\mathfrak{o}}'/\bar{\mathfrak{o}}'\bar{\eta} = \deg(\bar{\eta})_P = n_1 + \dots + n_g$.

We must now show that $\dim_k \bar{\mathfrak{o}}'/\bar{\mathfrak{o}} < \infty$. There exist $\omega_1, \dots, \omega_h \in \bar{\mathfrak{o}}'$ such that $\bar{\mathfrak{o}}' = \sum_{i=1}^{h} \bar{\mathfrak{o}}\omega_i$. Hence there exist elements $c \in \bar{\mathfrak{o}}$, $c \neq 0$, such that $c\bar{\mathfrak{o}}' \subset \bar{\mathfrak{o}}$. Let $\mathfrak{L}$ be the set of all such c including zero. It is easy to see that $\mathfrak{L}$ is an ideal in both $\bar{\mathfrak{o}}'$ and $\bar{\mathfrak{o}}$, and that $\mathfrak{L}$ is the largest set with this property. ($\mathfrak{L}$ is called the <u>conductor</u> of $\bar{\mathfrak{o}}'$ in $\bar{\mathfrak{o}}$). Since $\dim_k \bar{\mathfrak{o}}'/\mathfrak{L} < \infty$, we have, a fortiori, $\dim_k \bar{\mathfrak{o}}'/\bar{\mathfrak{o}} < \infty$.

<u>Cor. 7.9:</u> If $Y_1 \equiv Y_2$ and if Y_i has no common component with X, $i = 1, 2$, then $(Y_1.X) = (Y_2.X)$.

Proof: We can assume X is a prime cycle Γ and extend the result by linearity. Let $Y = Y_1 - Y_2 = (\eta)$, and let $P \in \Gamma$. Then $i(Y,\Gamma;P) = \deg(\bar{\eta})_P$, and $(Y.\Gamma) = \sum_{P\in\Gamma} i(Y,\Gamma;P) = \deg(\bar{\eta})$. But $\deg(\bar{\eta})$ is simply the sum of the orders of the zeros and poles of $\bar{\eta}$, which is zero.

<u>Def. 7.10:</u> If Y and X are any two cycles, then we define the <u>intersection number</u> $(Y.X)$ to be $(Y_1.X)$ where $Y_1 \equiv Y$ and Y_1, X have no common components.

By the preceding corollary this definition is independent of the choice of Y_1. The existence of a Y_1 follows from the fact there there exist $\xi \in k(V)$, $\xi \neq 0$, such that $v_\Gamma(\xi) = -m_\Gamma(Y)$ for all prime components of X. If we set $Y_1 = Y + (\xi)$, then Y_1 satisfies the requirements of Def. 7.10.

We call $(X.X) = (X^2)$ the <u>self-intersection number</u>. Clearly $(X+Y.X+Y) = (X^2) + 2(X.Y) + (Y^2)$ and $(nX)^2 = n^2(X^2)$. If X is a prime cycle and $\dim |X| \geq 1$, then $(X^2) \geq 0$ because dim $|X| \geq 1$ implies there exists $Y \geq 0$, such that $Y \equiv X$ and $Y \neq X$.

In addition to the divisorial cycles considered up to now, we shall also deal now with <u>zero-dimensional cycles</u> (briefly: <u>zero-cycles</u>): these are the elements of the free (additive) group generated by the points of F.

<u>Def. 7.11</u>: Let X and Y be cycles without common components. We define the <u>intersection cycle</u> of X and Y (denoted by $X.Y$) to be $\Sigma\, i(X,Y;P)P$.

Let $Z = \Sigma_P m_P P$ be a zero-cycle. Defining $\deg Z = \Sigma\, m_P$, we get $(X.Y) = \deg X.Y$.

Let Γ be a prime cycle and X a cycle such that Γ is not a component of X. Let $\eta^P = 0$ be a local equation of X at P where $P \in \Gamma$. Then $Tr_\Gamma \eta^P = \overline{\eta^P} \in k(\Gamma)$. We define $Tr_\Gamma X$ by $(Tr_\Gamma X)_P = \Sigma_P (\overline{\eta^P})_P$. Thus $X.\Gamma$ is a 0-dimensional cycle on F, and $Tr_\Gamma X$ is a 0-dimensional cycle on Γ.

Let L be a linear system on F and assume Γ is not a fixed component of L. Let $\bar{L} = \{Tr_\Gamma X | X \in L, \Gamma$ not a component of $X\}$. Then

(1) $\bar{L}$ is a linear system on Γ,

(2) if $\bar{B}$ is the fixed component of $\bar{L}$, then $\bar{L} - \bar{B} = \mathrm{Red}\, Tr_\Gamma L$.

We define $Tr_\Gamma L$ to be $\bar{L}$.

<u>Cor. 7.12</u>: $\dim L = \dim_\Gamma Tr_\Gamma L + \dim (L - \Gamma) + 1$.

Proof: Obvious (Theorem 5.8).

§8. Differentials

Let K/k be an algebraic function field of r variables where k is algebraically closed. We consider derivations of K/k, i.e., maps D of K into K satisfying

$$\begin{aligned} D(x+y) &= Dx + Dy, \\ D(xy) &= x\,Dy + y\,Dx, \\ Dc &= 0 \end{aligned}$$

for all x, y in K and all c in k. Let $\mathfrak{D}$ denote the set of all derivations of K/k; it is a vector space over K in virtue of the definition: $(\xi D)(x) = \xi(Dx)$ for all ξ and x in K.

We state (without proofs) two well-known facts:

(1) $\mathfrak{D}$ is an r-dimensional vector space over K.

(2) If $\{x_1, \ldots, x_r\}$ is a separating transcendence basis of K/k, then there exist r derivations $\partial/\partial x_i$ such that $\frac{\partial x_j}{\partial x_i} = \delta_{ij}$ $(i, j = 1, \ldots, r)$, and these derivations form a basis of $\mathfrak{D}$ over K. Conversely, if the derivations $\frac{\partial}{\partial x_i}$ exist, then $\{x_1, \ldots, x_r\}$ is a separating transcendence basis of K/k.

Let $\mathfrak{D}^*$ denote the dual space of $\mathfrak{D}$.

Def. 8.1: A differential of K/k of degree 1 is an element of $\mathfrak{D}^*$. A differential of degree q is a multilinear antisymmetric function on the direct product, $\mathfrak{D}^q$, of $\mathfrak{D}$ with itself q times with values in K.

If $t \in K$, we define dt, the differential of t, by the equation $dt(D) = Dt$ for all $D \in \mathfrak{D}$. If $t_1, \ldots, t_q \in K$, we define $dt_1 dt_2 \ldots dt_q$ by

$$(dt_1 \ldots dt_q)(D_1, \ldots, D_q) = \det\,(D_i(t_j)).$$

If $\{x_1, \ldots, x_r\}$ is a separating transcendence basis of K/k, then any differential ω_q, of degree q, can be written as

$$\omega_q = \sum_{i_1 < i_2 < \ldots < i_q} A_{i_1, \ldots, i_q} dx_{i_1} \ldots dx_{i_q}, \quad A_{i_1, \ldots, i_q} \in K.$$

For any differential ω_q of degree q we define the differential $d\omega_q$ of degree $q+1$ by $d\omega_q(D_1, \ldots, D_{q+1}) = \sum_{i=1}^{q+1} (-1)^{i-1} D_i(\omega_q(D_1, \ldots, \hat{D}_i, \ldots, D_{q+1}))$. ω_q is <u>closed</u> if $d\omega_q = 0$; ω_q is <u>exact</u> if there exists a differential ω_{q-1} such that $d\omega_{q-1} = \omega_q$.

<u>Theorem 8.2</u>: Let P be a rational simple point of an r-dimensional variety V/k. Let $\mathfrak{o} = \mathfrak{o}_P(V/k)$, $\mathfrak{m} = \mathfrak{m}_P(V/k)$ and let $t_1, \ldots, t_r$ be uniformizing parameters of P on V (i.e. $\mathfrak{m} = \sum_{i=1}^{r} \mathfrak{o}\, t_i$). Then $\{t_1, \ldots t_r\}$ is a separating transcendence basis of K/k, and $\frac{\partial}{\partial t_i} \mathfrak{o} \subset \mathfrak{o}$ for $i = 1, \ldots r$.

Proof: To prove this theorem we shall use the following well-known result (which we state without proof):

(3) If an extension $k(x)$ of a field k has no non-trivial derivation over k, it is separably algebraic over k, and conversely.

In view of (3) we must show that if D is a derivation of $K/k(t_1, \ldots, t_r)$ (hence D is trivial on $k(t_1, \ldots, t_r)$), then $D = 0$. Since K is the quotient field of $\mathfrak{o}$, we need only consider the effect of D on $\mathfrak{o}$. If $\xi \in \mathfrak{o}$, then, for any $i \geq 0$, there exists a polynomial $f_i(t_1, \ldots, t_r)$ in $k[t]$ such that $\xi - f_i \in \mathfrak{m}^{i+1}$. Since $\mathfrak{o}$ is the quotient ring of a finite integral domain, there exists an element A of $\mathfrak{o}$, $A \neq 0$, such that $(AD)(\mathfrak{o}) \subset \mathfrak{o}$. Replace D by AD. So we may assume that $D\mathfrak{o} \subset \mathfrak{o}$. It follows that $D\mathfrak{m}^{i+1} \subset \mathfrak{m}^i$. Since D is trivial on $k(t_1, \ldots, t_r)$, we have $D(\xi - f_i) = D\xi \in \mathfrak{m}^i$ for all i. Therefore $D\xi = 0$ because $\bigcap_{i=1}^{\infty} \mathfrak{m}^i = (0)$.

Let D be any of the $\partial/\partial t_i$. There exists an element A in $\mathfrak{o}$, $A \neq 0$, such that if we set $D' = AD$, then $D'\mathfrak{o} \subset \mathfrak{o}$. Then $D'\mathfrak{m}^{i+1} \subset \mathfrak{m}^i$; and $D't_j \equiv 0 \pmod{A}$ in $\mathfrak{o}$. This implies $D'k[t] \subset Ak[t] \subset A\mathfrak{o}$. Hence $D'\xi \in \bigcap_{i=1}^{\infty} \mathfrak{o}(A + \mathfrak{m}^i) = \mathfrak{o}A$. Therefore $D\xi \in \mathfrak{o}$.

Def. 8.3: A derivation D is regular at a given valuation v of K/k if $DR_v \subset R_v$ where R_v is the valuation ring of v.

It is clear that the derivations which are regular at v form a module over R_v.

Def. 8.4: A differential ω_q is regular at v if $\omega_q(D_1, \ldots, D_q) \in R_v$ for all $D_1, \ldots, D_q$ which are regular at v. A differential ω_q is a regular differential of K/k if it is regular at each valuation v of K/k.

Notation: $D_v = \{D \in \mathcal{D} \mid D \text{ regular at } v\}$.

Using this notation, we have $\omega_q(\mathcal{D}_v^q) \subset R_v$.

Let V be a variety such that $k(V) = K$.

Def. 8.5: Let P be a point of V and D a derivation of K/k. Then D is regular at P if $D\bar{\mathfrak{o}}_P \subset \bar{\mathfrak{o}}_P$ where $\bar{\mathfrak{o}}_P$ is the integral closure of $\mathfrak{o}_P$ in K.

Def. 8.6: A differential ω_q is regular at P if $\omega_q(D_1, \ldots, D_q) \in \bar{\mathfrak{o}}_P$ for all $D_1, \ldots, D_q$ which are regular at P. A differential ω_q is regular on V if it is regular at each point of V.

Prop. 8.7: A differential ω_q is regular on V if either one of the following conditions is satisfied:

(1) ω_q is regular at each rational point of V.

(2) ω_q is regular at each irreducible $(r-1)$-dimensional subvariety Γ of V.

Proof: Assume (1) and let P be any point of V. Let P' be a specialization of P over k; then $\bar{\mathfrak{o}}_P$ is a ring of fractions of $\bar{\mathfrak{o}}_{P'}$, i.e. $\bar{\mathfrak{o}}_P = (\bar{\mathfrak{o}}_{P'})_S$ where S is a multiplicatively closed subset of $\bar{\mathfrak{o}}_{P'}$. Let D be a derivation which is regular at P. There exists an element $f \in S$ such that fD is regular at P'. Hence if $D_1, \ldots, D_q$

are regular at P, we can find $f \in S$ such that fD_i is regular at P' for $i = 1, \ldots, q$. Now every point P has an algebraic specialization P'; and since k is algebraically closed, P' is rational. Since ω_q is regular at P', we have $\omega_q(fD_1, \ldots, fD_q) = f^q \omega_q(D_1, \ldots, D_q) \in \bar{\mathfrak{o}}_{P'}$. Hence $\omega_q(D_1, \ldots, D_q) \in \bar{\mathfrak{o}}_P$, showing that ω_q is regular at P.

Assume (2). Let $P \in V$ and assume $D_1, \ldots, D_q$ are regular at P. Then $D_i \bar{\mathfrak{o}}_P \subseteq \bar{\mathfrak{o}}_P$, $i = 1, \ldots, q$. We note that if $\mathfrak{o}$ is any subring of $k(V)$ and if $D\mathfrak{o} \subseteq \mathfrak{o}$, then $D\mathfrak{o}_S \subseteq \mathfrak{o}_S$ where S is a multiplicatively closed subset of $\mathfrak{o}$; for if $f = g/h \in \mathfrak{o}_S$, $g, h \in \mathfrak{o}$, $h \in S$, then $Df = (1/h^2)(hDg - g\,Dh) \in \mathfrak{o}_S$. Therefore $D_1, \ldots, D_q$ are regular at all Γ which contain P, and so $\omega_q(D_1, \ldots, D_q) \in \bar{\mathfrak{o}}_\Gamma$ for all such Γ. Let $A = \omega_q(D_1, \ldots, D_q)$ and assume $A \notin \bar{\mathfrak{o}}_P$. We know $\bar{\mathfrak{o}}_P = \bigcap_{\mathfrak{p}} (\mathfrak{o}_P)_{\mathfrak{p}}$ where $\mathfrak{p}$ runs through the minimal (rank 1) prime ideals of $\mathfrak{o}_P$. Hence there exists a $\mathfrak{p}$ such that $A \notin (\mathfrak{o}_P)_{\mathfrak{p}}$. Let Γ be the center of the valuation $\mathfrak{p}$. Then $A \notin \bar{\mathfrak{o}}_\Gamma$ which is a contradiction. Hence $A \in \bar{\mathfrak{o}}_P$ and ω_q is regular at P.

Now, let Γ be any irreducible $(r-1)$-dimensional subvariety of V such that $P \in \Gamma$. It is immediately seen that the integral closure $\bar{\mathfrak{o}}_\Gamma$ of the local ring $\mathfrak{o}_\Gamma(V/k)$ is the ring of quotients of $\bar{\mathfrak{o}}_P$ with respect to the multiplicative system $\mathfrak{o}_P - \mathfrak{p}$, where $\mathfrak{p}$ is the prime ideal of Γ in $\mathfrak{o}_P$. Since $D_i \bar{\mathfrak{o}}_P \subset \bar{\mathfrak{o}}_P$, it follows that $D_i \bar{\mathfrak{o}}_\Gamma \subset \bar{\mathfrak{o}}_\Gamma$, i.e., the D_i are regular at Γ. Hence $\omega_q(D_1, \ldots, D_q) \in \bar{\mathfrak{o}}_\Gamma$. Since the intersection of all the rings $\bar{\mathfrak{o}}_\Gamma$ is $\bar{\mathfrak{o}}_P$, ω_q is regular at P.

<u>Theorem 8.8</u>: If ω_q is a regular differential of the field K/k, then ω_q is regular on V; and conversely, provided V is non-singular.

Remark: We shall assume $q=1$. The proof for larger values of q is the same (but with more indices).

Proof: Assume ω is a regular differential of K/k. Let $\Gamma \subset V$ here $\dim \Gamma = r-1$, let $\mathfrak{o} = \mathfrak{o}_{\Gamma}(V/k)$ and let $\bar{\mathfrak{o}}$ be the integral closure of $\mathfrak{o}$. There are a finite number of prime divisors $v_1, \ldots, v_h$ of K/k with center Γ. There exist exactly h prime ideals $\mathfrak{p}_1, \ldots, \mathfrak{p}_h$ in $\bar{\mathfrak{o}}$ all of which are maximal, and $\bar{\mathfrak{o}}_{\mathfrak{p}_i} = R_{v_i}$, $\bar{\mathfrak{o}} = \bigcap_{i=1}^{h} R_{v_i}$. Let D be a derivation which is regular at Γ. Then $D\bar{\mathfrak{o}} \subset \bar{\mathfrak{o}}$ and so $DR_{v_i} \subset R_{v_i}$. Hence $\omega(D) \in \bigcap_{i=1}^{h} R_{v_i} = \bar{\mathfrak{o}}$. Therefore ω is regular at Γ; and Prop. 8.7(2) shows that ω is regular on V.

Now assume V is non-singular, and ω is regular on V. Let v be any valuation of K/k, and let W be the center of v on V (hence W is an irreducible subvariety of V). Fix a rational point P on W and uniformizing parameters $t_1, \ldots, t_r$ of P on V. We can write $\omega = \sum_{i=1}^{r} A_i dt_i$ where the A_i are in K. Indeed, the regularity of ω implies the A_i are in $\mathfrak{o}_P$ in view of Theorem 8.2. Let D be any derivation which is regular at v. We can write $D = \sum_{i=1}^{r} B_i \frac{\partial}{\partial t_i}$. Since $t_i \in \mathfrak{o}_P \subset \mathfrak{o}_W \subset R_v$, we have $Dt_i = B_i \in R_v$. Hence, since $\omega(\frac{\partial}{\partial t_i}) = A_i$, $\omega(D) = \sum_{i=1}^{r} B_i A_i \in R_v$. Therefore ω is regular at v.

§9. The canonical system on a variety V.

Def. 9.1: If W is a simple irreducible subvariety of V/k, and $\xi_1, \ldots, \xi_r \in \mathcal{O} = \mathcal{O}_W(V/k)$ ($r = \dim V$), then the ξ_i are uniformizing coordinates of W if

(a) $\{\xi_1, \ldots, \xi_r\}$ is a separating transcendence basis of K/k

(b) $\frac{\partial}{\partial \xi_i} \mathcal{O} \subset \mathcal{O}$

Let W be any simple irreducible subvariety of V/k. Then it is easily seen that there exist sets of uniformizing coordinates. For, let $P \in W$ where P is a rational simple point of V. Let $\{t_1, \ldots, t_r\}$ be a set of uniformizing parameters of P. Then by Theorem 8.2 we know (a) holds; and since $\frac{\partial}{\partial t_i}$ are regular at P, they are regular at W.

An alternative method is as follows: let V_a be an affine representative of V such that W has an affine representative W_a in V_a, i.e., $W \not\subset V - V_a$. Let $k[V_a] = k[x_1, \ldots, x_n]$, $x_i \in \mathcal{O}$; and let $\{f_1(X), \ldots, f_N(X)\}$ be a basis of $I(V_a)$. Consider the Jacobian $\frac{\partial(f_1, \ldots, f_N)}{\partial(X_1, \ldots, X_n)}$. W is simple on V if and only if this matrix has rank $n-r$ on W, hence we can assume $\frac{\partial(f_1, \ldots, f_{n-r})}{\partial(x_{r+1}, \ldots, x_n)} \neq 0$ on W.

Thus $\{x_1, \ldots, x_r\}$ is easily seen to be a set of uniformizing coordinates for W.

Prop. 9.2: If $\xi_1, \ldots, \xi_r$ are uniformizing coordinates of W and if $\eta_1, \ldots, \eta_r \in \mathcal{O}$, then $\eta_1, \ldots, \eta_r$ are uniformizing coordinates of W if and only if $\frac{\partial(\eta_1, \ldots, \eta_r)}{\partial(\xi_1, \ldots, \xi_r)} \notin \mathfrak{m}_W$.

Proof: Assume the η_i are uniformizing coordinates. Then $\frac{\partial(\eta)}{\partial(\xi)}\frac{\partial(\xi)}{\partial(\eta)} = 1$, and (b) implies $\frac{\partial(\eta)}{\partial(\xi)}\ \frac{\partial(\xi)}{\partial(\eta)} \in \mathcal{O}$. Hence $\frac{\partial(\eta)}{\partial(\xi)}$ is a unit in $\mathcal{O}$.

Let $D_i = \Sigma_{j=1}^r A_{ij} \frac{\partial}{\partial \xi_j}$, $i = 1, \ldots, r$, $A_{ij} \in K$. We want to choose the A_{ij} so that the system of linear equations $D_i \eta_j = \delta_{ij}$ holds. Regarding these as equations for the A_{ij}, the determinant is $\frac{\partial(\eta)}{\partial(\xi)}$ which is a unit in $\mathcal{O}$. Hence we can find such $A_{ij} \in \mathcal{O}$. This proves (a) of Def. 9.1. Clearly $D_i \mathcal{O} \subset \mathcal{O}$ since $\frac{\partial}{\partial \xi_i}$; $\mathcal{O} \subset \mathcal{O}$.

Let $\mathcal{P}$ be a prime divisor of K/k of the first kind with respect to V, and let W be the center of $\mathcal{P}$ on V (hence $\dim W = r-1$). Choose an affine representative V_a of V such that W has an affine representative W_a contained in V_a. Let $R = k[V_a]$, and let $\bar{R} = k[z_1, \ldots, z_m]$ be the integral closure of R in its quotient field. Let $\bar{V}$ be the locus of $(z_1, \ldots, z_m)$ over k. Then $\bar{V}$ is an affine variety, and $k(\bar{V}) = k(V)$ (because R and $\bar{R}$ have the same quotient field). We call $\bar{V}$ a <u>normalization</u> of V_a. We note that $\bar{V}$ is determined only up to a biregular transformation since we only know its coordinate ring.

Let $\bar{W}$ be the center of $\mathcal{P}$ on $\bar{V}$. Clearly $\dim \bar{W} = r-1$ because $\bar{W}$ is represented by a minimal prime ideal in $\bar{R}$. Hence the valuation ring of $\mathcal{P}$ is $\mathcal{O}_{\bar{W}}(\bar{V}/k)$. By a <u>set of uniformizing coordinates of</u> $\mathcal{P}$ we mean any set of uniformizing coordinates of $\bar{W}$ on $\bar{V}$.

Let ω_r be an r-fold differential of $k(V)/k$, $\omega_r \neq 0$; let $\mathcal{P}$ be a prime divisor of the first kind with respect to V; and let $\xi_1, \ldots, \xi_r$ be uniformizing coordinates of $\mathcal{P}$. Then $d\xi_1 \ldots d\xi_r \neq 0$, and we can write $\omega_r = A d\xi_1 \ldots d\xi_r$ where $A \in K$.

Let $v_{\mathfrak{p}}$ be the valuation defined by $\mathfrak{p}$. We define the order of ω_r at $\mathfrak{p}$ (notation: $v_{\mathfrak{p}}(\omega_r)$) to be $v_{\mathfrak{p}}(A)$. To see that $v_{\mathfrak{p}}(\omega_r)$ is independent of the choice of uniformizing coordinates, let $\eta_1, \ldots, \eta_r$ be another set. Then $\omega_r = B d\eta_1 \ldots d\eta_r$ and $B = A \left| \frac{\partial(\xi)}{\partial(\eta)} \right|$. Hence $v_{\mathfrak{p}}(A) = v_{\mathfrak{p}}(B)$ since $v_{\mathfrak{p}}\left(\left|\frac{\partial(\xi)}{\partial(\eta)}\right|\right) = 0$ (by Prop. 9.2, this Jacobian determinant is a unit in the quotient ring $\mathcal{O}_{\overline{W}}(\overline{V}/k)$).

Prop. 9.3: Let ω_r be an r-fold differential of $k(V)/k$, $\omega_r \neq 0$. Then $v_{\mathfrak{p}}(\omega_r) = 0$ for all but a finite number of prime divisors $\mathfrak{p}$ of the first kind with respect to V.

Proof: Let $V_a = V - (V \cap H)$ where H is a hyperplane such that $V \not\subset H$. Since $V \cap H$ contains only a finite number of prime cycles, we may replace in the proof the variety V by its affine representative V_a.

Let $R = k[V_a] = k[x_1, \ldots x_n]$ where we can assume that $\{x_1, \ldots, x_r\}$ is a separating transcendence basis of $k(V)/k$. Let $\overline{R} = k[z_1, \ldots, z_m]$ be the integral closure of R, and consider $\partial z_\nu / \partial x_i$, $\nu = 1, \ldots, m$; $i = 1, \ldots, r$. There is only a finite number of prime divisors which are poles of the $\partial z_\nu / \partial x_i$ and so we can discard them. Hence $\partial z_\nu / \partial x_i$ ($\nu = 1, \ldots, m$; $i = 1, \ldots, r$) are finite at all the remaining prime divisors $\mathfrak{p}$. Therefore $\partial z_\nu / \partial x_i \in R_{\mathfrak{p}}$ for all i and all ν. This implies that $\frac{\partial}{\partial x_i} \overline{R} \subset R_{\mathfrak{p}}$, whence $\frac{\partial}{\partial x_i} R_{\mathfrak{p}} \subset R_{\mathfrak{p}}$ for $i = 1, \ldots, r$; and so $x_1, \ldots, x_r$ are uniformizing coordinates of $\mathfrak{p}$. Hence, if $\omega_r = A dx_1 \ldots dx_r$, ($A \in K$), then for all $\mathfrak{p}$ outside the two finite sets we have excluded, we have $v_{\mathfrak{p}}(\omega_r) = v_{\mathfrak{p}}(A)$ which proves the proposition.

Def. 9.4: The divisor $(\omega_r) = \Sigma\, v_{\mathfrak{p}}(\omega_r) \cdot \mathfrak{p}$, $\omega_r \neq 0$, where $\mathfrak{p}$ runs the set of all the prime divisors of the first kind with respect to V, is called the divisor of the differential ω_r.

A <u>canonical divisor</u> on V is the divisor of an r-fold differential ω_r, $\omega_r \neq 0$. Let ω_r, ω'_r be r-fold differentials, $\omega_r \neq 0, \omega'_r \neq 0$. Then $\omega'_r = B\omega_r$ where $B \in k(V)$. Hence $(\omega'_r) = (B) + (\omega_r)$, i.e. (ω_r) and (ω'_r) are linearly equivalent. Thus all canonical divisors belong to one and the same divisor class, and, furthermore, it is clear that any divisor linearly equivalent to a canonical divisor is canonical. This divisor class is called the <u>canonical divisor class</u>. If K is any canonical divisor, then $|K|$ is the <u>canonical system on</u> V.

An immediate consequence of our definitions is

<u>Prop. 9.5</u>: A differential ω_r is regular on V if and only if $(\omega_r) \geq 0$.

Let ω_r be a regular differential, and let $K = (\omega_r)$. By definition, K is an effective divisor. Conversely, if we start with an effective canonical divisor K, then ω_r is determined to within a non-zero constant factor in k. Let L be the vector space (over k) of the regular differentials on V, of degree r. We have a mapping $L - \{0\} \to |K|$, where $\omega_r \to (\omega_r)$, $\omega_r \neq 0$. Thus we can consider $|K|$ as a projective space. Since $\dim |K| < \infty$, we have $\dim L < \infty$: in fact, $\dim L = 1 + \dim |K|$. We denote $\dim L$ by $p_g(V)$, and call $p_g(V)$ the <u>geometric genus</u> of the variety V. Hence V carries exactly $p_g(V)$ linearly independent differentials of degree r which are regular on V. Clearly, $p_g(V)$ is always ≥ 0.

If V is biregularly equivalent to V', the $p_g(V) = p_g(V')$. This is not necessarily true for birationally equivalent varieties. However, if V is non-singular and $k(V) = k(V')$, the $p_g(V) \leq p_g(V')$, because any differential of degree r which is regular on V is regular on $k(V)$ and hence on V'. This proves

Prop. 9.6: Any two non-singular models V, V' of a given function field Σ ($= k(V) = k(V')$) have the same geometric genus.

Let Σ/k be a function field where t.d. $\Sigma/k = r$. We define the geometric genus of Σ/k, $p_g(\Sigma/k)$, to be the minimum value of $p_g(V)$ where V is any projective model of Σ/k.

Theorem 9.7: If V/k is a normal variety, then the differentials of a given degree q which are regular on V form a finite-dimensional vector space over k.

Proof: We shall assume $q = 1$ since the proof is the same for any q.

Fix a separating transcendence basis $\{\xi_1, \ldots, \xi_r\}$ of $k(V)/k$, and let ω be any differential of degree 1 which is regular on V. Then we can write $\omega = A_1 d\xi_1 + \ldots + A_r d\xi_r$. Let Γ be a prime divisorial cycle.

Case 1: Assume the ξ_i are uniformizing coordinates at Γ. Since ω is regular (in particular, regular at Γ) the A_i must be regular at Γ. Hence $v_\Gamma(A_i) \geq 0$ for $i = 1, \ldots, r$.

Case 2: Assume the ξ_i are not uniformizing coordinates at Γ. Let $\eta_1, \ldots, \eta_r$ be uniformizing coordinates at Γ. Then

$$d\xi_1 \ldots d\xi_r = \left|\frac{\partial(\xi_1, \ldots, \xi_r)}{\partial(\eta_1, \ldots, \eta_r)}\right| d\eta_1 \ldots d\eta_r.$$

Since the η_i are uniformizing coordinates, we have

$$v_\Gamma(d\xi_1 \ldots d\xi_r) = v_\Gamma\left(\frac{\partial(\xi)}{\partial(\eta)}\right) = \text{coefficient of } \Gamma \text{ in the divisor } (d\xi_1 \ldots d\xi_r).$$

Since the ξ_i are not uniformizing coordinates of Γ, either some ξ_i is infinite at Γ or Γ is a component of $(d\xi_1 \ldots d\xi_r)$. Thus there are only a finite number of prime divisorial cycles Γ such that the ξ_i are not uniformizing coordinates of Γ. Let $\mathfrak{A}$ be the set of these cycles Γ.

Let $\Gamma \in \mathfrak{A}$, and fix uniformizing coordinates $\eta_1, \ldots \eta_r$ at Γ. Then $\omega = B_1 d\eta_1 + \ldots + B_r d\eta_r$ where $B_1, \ldots, B_r \in \mathfrak{o}_\Gamma$. Then $A_i = \sum_{j=1}^{r} B_j \frac{\partial \eta_j}{\partial \xi_i}$, $i = 1, \ldots, r$. This gives us a lower bound for the orders of the A_i, namely

$$v_\Gamma(A_i) \geq \min_{\Gamma \in \mathfrak{A}} \left\{ \{\ldots, v_\Gamma(\ldots, \frac{\partial \eta_\nu}{\partial \xi_\mu}, \ldots), \ldots\}, 0 \right\}. \quad (*)$$

Denote the right-hand side of (*) by $s(\Gamma)$. Let $\mathfrak{A} = \{\Gamma_1, \ldots, \Gamma_h\}$, and let $s_i = s(\Gamma_i)$. Define $Z = -\sum_{i=1}^{h} s_i \Gamma_i$. We have shown that $(A_i) + Z \geq 0$. Hence each A_i varies in a finite-dimensional vector space over k, and hence so does ω.

§10. Trace of a differential.

Let W be a subvariety of V; and let $\mathfrak{o} = \mathfrak{o}_W(V/k)$, $\mathfrak{m} = \mathfrak{m}_W(V/k)$, and $\Sigma = k(V)$. Let $\mathcal{D}_W$ be the set of all derivations which are regular at W; then $\mathcal{D}_W$ is an $\mathfrak{o}$-module. Let $\mathcal{D}_W^\circ = \{D \in \mathcal{D}_W \mid D\mathfrak{m} \subset \mathfrak{m}\}$.

Def. 10.1: If $D \in \mathcal{D}_W^\circ$ we define the trace $\mathrm{Tr}_W D$ of D on W to be the derivation $\bar{D}$ of $k(W)$ such that $\bar{D}(\bar{\eta}) = \mathrm{Tr}_W(D\eta)$ where $\eta \in \mathfrak{o}$, $\bar{\eta} = \mathrm{Tr}_W \eta$ ($\bar{\eta} \in k(W) = \mathfrak{o}/\mathfrak{m}$).

If $\eta \in \mathfrak{m}$, then $D\eta \in \mathfrak{m}$ and $\mathrm{Tr}_W \eta = 0$. Hence $\bar{D}$ is well-defined. It is immediately seen that $\bar{D}$ is indeed a derivation of $k(W)/k$.

Prop. 10.2: If W is a simple subvariety of V, then every derivation of $k(W)/k$ is the trace of some derivation in $\mathcal{D}_W^\circ$.

Proof: Let $s = \dim W$. If $s = 0$, then W is a point and the result is trivial. Hence we may assume $s > 0$. Fix uniformizing coordinates $\xi_1, \ldots, \xi_r$ and uniformizing parameters $t_1, \ldots, t_{r-s}$ of W on

V/k (hence $\mathfrak{m} = \sum_{i=1}^{r-s} \mathfrak{O} t_i$). The quantities $\frac{\partial t_i}{\partial \xi_\nu}$, $i = 1, \ldots, r-s$; $\nu = 1, \ldots, r$, belong to $\mathfrak{O}$. Let $Tr_W \frac{\partial t_i}{\partial \xi_\nu} = \bar{z}_{i\nu} \in k(W)$, $i = 1, \ldots, r-s$; $\nu = 1, \ldots, r$; and let $\bar{z}_i = (\bar{z}_{i1}, \ldots, \bar{z}_{ir})$, $i = 1, \ldots, r-s$. Let $r-s-\sigma$ be the number of the vectors $\bar{z}_i$ which are linearly independent over $k(W)$ (hence $\sigma \geq 0$). Let $e_1 = (1,0, \ldots, 0)$, $e_2 = (0,1,0,\ldots, 0)$, $\ldots$, $e_r = (0, \ldots, 0,1)$. We can assume that $e_1, \ldots, e_{s+\sigma}$ span a space complementary to that spanned by the $\bar{z}_i$.

We can find $L_\alpha = \sum_{\nu=1}^{r} \bar{a}_{\alpha\nu} X_\nu$, $\alpha = 1, \ldots, s + \sigma$, $\bar{a}_{\alpha\nu} \in k(W)$, such that

$$L_\alpha(e_\beta) = \delta_{\alpha\beta}, \quad \beta = 1, \ldots, s + \sigma$$

$$L_\alpha(\bar{z}_i) = 0, \quad i = 1, \ldots, r - s$$

Let $a_{\alpha\nu} \in \mathfrak{O}$ have W-trace $\bar{a}_{\alpha\nu}$, and consider the derivations $D^\circ_\alpha = \sum_{\nu=1}^{r} a_{\alpha\nu} \frac{\partial}{\partial \xi_\nu}$, $\alpha = 1, \ldots, s + \sigma$. Clearly $D^\circ_\alpha \in \mathfrak{D}_W$ for all α. We have $D^\circ_\alpha t_i = \sum_{\nu=1}^{r} a_{\alpha\nu} \frac{\partial t_i}{\partial \xi_\nu} \in \mathfrak{O}$. Hence $Tr_W(D^\circ_\alpha t_i) = \sum_{\nu=1}^{r} \bar{a}_{\alpha\nu} \bar{z}_{i\nu} = L_\alpha(\bar{z}_i) = 0$. Therefore $D^\circ_\alpha t_i \in \mathfrak{m}$ for $\alpha = 1, \ldots, s + \sigma$; $i = 1, \ldots, r - s$; and so $D^\circ_\alpha \mathfrak{m} \subset \mathfrak{m}$. This shows $D^\circ_1, \ldots, D^\circ_{s+\sigma} \in \mathfrak{D}^\circ_W$.

Now consider the traces $\bar{D}^\circ_\alpha$ of the D°_α on W. We have $D^\circ_\alpha \xi_\beta \equiv \delta_{\alpha\beta} \pmod{\mathfrak{m}}$, i.e., $\bar{D}^\circ_\alpha \bar{\xi}_\beta = \delta_{\alpha\beta}$, $\beta = 1, \ldots, s + \sigma$, where $\bar{\xi}_\beta = Tr_W \xi_\beta$. Hence $D^\circ_1, \ldots, D^\circ_{s+\sigma}$ are linearly independent over $k(W)$. Cpnsequently $\sigma = 0$. Thus we have shown that <u>if</u> $\mathfrak{m} = \sum_{i=1}^{r-s} \mathfrak{O} t_i$, <u>then</u> $\left\| \frac{\partial t_i}{\partial \xi_\nu} \right\|$ <u>has (maximum possible) rank</u> $r-s$ <u>on</u> W.

It follows that we can replace $r-s$ of the ξ's by the t's. We may therefore assume that $\xi_{s+i} = t_i$, $i = 1, \ldots, r - s$. Let

$D_\alpha = \frac{\partial}{\partial \xi_\alpha}$, $\alpha = 1, \ldots, s$. Then $D_\alpha t_i = 0$ for $i = 1, \ldots, r - s$; $\alpha = 1, \ldots, s$; and since the t_i form a basis of $\mathfrak{m}$, we have $D_\alpha \mathfrak{m} \subset \mathfrak{m}$, $\alpha = 1, \ldots, s$. Therefore $D_1, \ldots, D_s \in \mathfrak{D}^\circ_W$. Let $\bar{D}_\alpha = \mathrm{Tr}_W D_\alpha$ and let $\bar{\xi}_\alpha = \mathrm{Tr}_W \xi_\alpha$. Then, by definition, $\bar{D}_\alpha \bar{\xi}_\beta = \delta_{\alpha\beta}$ for $\alpha, \beta = 1, \ldots, s$. Therefore the $\bar{D}_\alpha$ form a basis of the space $\mathfrak{D}(W)$ of all derivations of $k(W)/k$, and this proves the proposition.

<u>Prop. 10.3</u>: If W is a simple subvariety of V/k of dimension s, then

(a) $\mathfrak{D}_W$ is a free r-dimensional $\mathfrak{O}$-module ($\mathfrak{O} = \mathfrak{O}_W(V/k)$).

(b) $\mathfrak{D}^\circ_W \supset \mathfrak{m}\mathfrak{D}_W$ and

(c) $\mathfrak{D}^\circ_W / \mathfrak{m}\mathfrak{D}_W$ is a free s-dimensional $\mathfrak{O}/\mathfrak{m}$-module.

Proof: Let $\xi_1, \ldots, \xi_r$ be uniformizing coordinates of W. Then $D \in \mathfrak{D}_W$ if, and only if, $D = \sum_{i=1}^{r} A_i \frac{\partial}{\partial \xi_i}$, with $A_i \in \mathfrak{O}$ since $D\xi_i = A_i$. Therefore $\left\{\frac{\partial}{\partial \xi_1}, \ldots, \frac{\partial}{\partial \xi_r}\right\}$ form an $\mathfrak{O}$-basis of $\mathfrak{D}_W$, and the module is free because the $\frac{\partial}{\partial \xi_i}$ are linearly independent over $k(V)$. This proves (a).

(b) is obvious.

Let $\{\xi_1, \ldots, \xi_r\}$ be chosen so that $\xi_{s+i} = t_i$, $i = 1, \ldots, r-s$, where the t_i are uniformizing parameters of W. Then, if $D \in \mathfrak{D}_W$, we can write $D = \sum_{i=1}^{s} A_i \frac{\partial}{\partial \xi_i} + \sum_{j=1}^{r-s} B_j \frac{\partial}{\partial t_j}$. Then $D\mathfrak{O} \subset \mathfrak{O}$; and $D\mathfrak{m} \subset \mathfrak{m}$ if, and only if,

(1) $A_i, B_j \in \mathfrak{O}$, $i = 1, \ldots, s$; $j = 1, \ldots, r - s$; and

(2) $Dt_j \subset \mathfrak{m}$.

(1) and (2) are equivalent to $B_j \in \mathfrak{m}$, $j = 1, \ldots, r - s$. This proves (c).

We now give a characterization of uniformizing coordinates:

<u>Prop. 10.4</u>: Let W be a simple s-dimensional subvariety of V/k, and let $\xi_1, \ldots, \xi_r \in \mathfrak{O} = \mathfrak{O}_W(V/k)$. Then the ξ_i are uniformizing coordinates of W on V if, and only if, the following two

conditions are satisfied:

(a) $k(W)$ is a separable algebraic extension of $k(\bar{\xi}_1, \ldots, \bar{\xi}_r)$ where $\bar{\xi}_i = Tr_W \xi_i$, $i = 1, \ldots, r$.

(b) The ring $k[\xi_1, \ldots, \xi_r]$ contains a set of uniformizing parameters of W on V/k.

Furthermore, (b) is equivalent to

(b') Let $\mathfrak{m} = \mathfrak{m}_W(V/k)$. Then $\mathfrak{m} \cap k[\xi]$ contains a basis of $\mathfrak{m}$.

Proof: Assume (a) and (b). Fix uniformizing coordinates $\eta_1, \ldots, \eta_r$ of W on V/k. By virtue of Prop. 9.2 it suffices to show that $\left|\frac{\partial(\xi)}{\partial(\eta)}\right| \notin \mathfrak{m}$. Assume $\left|\frac{\partial(\xi)}{\partial(\eta)}\right| \in \mathfrak{m}$. Then we can find $A_1, \ldots, A_r \in \mathcal{O}$, where not all the A_i are in $\mathfrak{m}$, say $A_1 \notin \mathfrak{m}$, such that $\sum_{j=1}^{r} A_j \frac{\partial \xi_i}{\partial \eta_j} \in \mathfrak{m}$. Let $D = \sum_{j=1}^{r} A_j \frac{\partial}{\partial \eta_j}$.

Since the η_j are uniformizing coordinates, we have $D \in \mathcal{D}_W$, i.e., D is regular on W. Since $D\xi_i \in \mathfrak{m}$, $i = 1, \ldots, r$, it follows that $D(k[\xi]) \in \mathfrak{m}$. Hence, by (b), we have $D\mathfrak{m} \subset \mathfrak{m}$. Therefore $D \in \mathcal{D}_W^{\circ}$, and so D has a trace $\bar{D} = Tr_W D$. $D\xi_i \in \mathfrak{m}$ implies $\bar{D}\bar{\xi}_i = 0$. This shows $\bar{D}$ is trivial on $k(\bar{\xi}_1, \ldots, \bar{\xi}_r)$ and hence $\bar{D}$ is trivial on $k(W)$ by (a). On the other hand, $D\eta_1 = A_1$, hence $\bar{D}\bar{\eta}_1 = \bar{A}_1 \neq 0$ and this is a contradiction. This shows that the ξ_i are uniformizing coordinates of W on V.

Now assume $\xi_1, \ldots, \xi_r$ are uniformizing coordinates of W. To show (a) we need only prove that if $\bar{D}$ is a derivation of $k(W)/k$ such that $\bar{D} = 0$ on $k(\bar{\xi})$, then $\bar{D} = 0$ on $k(W)$. Let $\bar{D} = Tr_W D$, $D \in \mathcal{D}_W^{\circ}$. Then $D\xi_i \in \mathfrak{m}$. Let $A_i = D\xi_i$; then $D = \sum_{i=1}^{r} A_i \frac{\partial}{\partial \xi_i}$. Since $A_i \in \mathfrak{m}$, we have $D\mathcal{O} \subset \mathfrak{m}$ which shows that $\bar{D} = 0$ on $k(W)$.

Since (a) holds, we can assume $\{\bar{\xi}_1, \ldots, \bar{\xi}_s\}$ is a separating transcendence basis of $k(W)/k$. Let $f_i(\bar{\xi}_1, \ldots, \bar{\xi}_{s+i-1}, X)$ be the minimal polynomial of $\bar{\xi}_{s+i}$ over $k(\bar{\xi}_1, \ldots, \bar{\xi}_{s+i-1})$ for $i = 1, \ldots, r-s$. By clearing denominators we may assume that f_i is a polynomaal in all its $s+i$ arguments (of course, f_i now need not be monic).

Let $t_i = f(\xi_1, \ldots, \xi_{s+i-1}, \xi_{s+i})$. Then clearly $t_i \in k[\xi]$; and, since $f_i(\bar{\xi}_1, \ldots, \bar{\xi}_{s+i}) = 0$, we have $t_i \in k[\xi] \cap \mathfrak{m}$, $i = 1, \ldots, r-s$. We shall show that $t_1, \ldots, t_{r-s}$ are uniformizing parameters of W. More precisely, if $\zeta = A_1 t_1 + \ldots + A_{r-s} t_{r-s} \in \mathfrak{m}^2$, $A_i \in \mathfrak{o}$, we shall show that $A_i \in \mathfrak{m}$ for $i = 1, \ldots, r-s$ (this means that the $\mathfrak{m}^2$-residues of the t_i are linearly independent over $\mathfrak{o}/\mathfrak{m}$). We know $\frac{\partial \zeta}{\partial \xi_r} \in \mathfrak{m}$. Hence $\frac{\partial \zeta}{\partial \xi_r} - \sum_{j=1}^{r-s} \frac{\partial A_j}{\partial \xi_r} t_j = A_{r-s} \frac{\partial f_{r-s}(\xi_1, \ldots, \xi_r)}{\partial \xi_r} \in \mathfrak{m}$. Since $k(W)$ is separable algebraic over $k(\bar{\xi}_1, \ldots, \bar{\xi}_s)$ we have $\frac{\partial f_{r-s}(\xi_1, \ldots, \xi_r)}{\partial \xi_r} \notin \mathfrak{m}$

Therefore $A_{r-s} \in \mathfrak{m}$. In a similar fashion we see that $A_i \in \mathfrak{m}$ for $i = 1, \ldots, r-s$.

Special case. Let $s = r - 1$, i.e., let W be a prime divisorial cycle. Then t.d. $k(W)/k = r - 1$. Let $\{\xi_1, \ldots, \xi_r\}$ be a set of uniformizing coordinates of W. Then among the $\bar{\xi}_i$, $i = 1, \ldots, r$, are $r - 1$ algebraically independent elements. Thus all elements of $k[\xi] \cap \mathfrak{m}$ are multiples of one irreducible polynomial $f(\xi_1, \ldots, \xi_r)$ which we denote by t. Since $k[\xi] \cap \mathfrak{m}$ contains a basis of $\mathfrak{m}$, the set $\{t\}$ must be a set of uniformizing parameters, i.e., $\mathfrak{m}$ is a principal ideal. Hence $v_W(t) = 1$.

Let W be a simple s-dimensional subvariety of V, and let $\{\xi_1, \ldots, \xi_r\}$ be a set of uniformizing coordinates of W such that

$\{\xi_{s+1}, \ldots, \xi_r\}$ is a set of uniformizing parameters of W. Then we can write $\mathcal{D}^\circ_W = \sum_{i=1}^{s} \mathcal{O} \frac{\partial}{\partial \xi_i} + m\mathcal{D}_W$. We have seen that $\mathcal{D}^\circ_W / m\mathcal{D}_W$ is an $\mathcal{O}/m$-module, i.e., a k(W)-module. By Prop. 10.2 and Prop. 10.3(c), we see that $\mathcal{D}^\circ_W / m\mathcal{D}_W$ can be identified with the space $\mathcal{D}(W)$ of derivations of k(W)/k.

Let ω_q be a q-fold differential, and <u>assume</u> ω_q <u>is regular at</u> W. Then we have the following:

<u>Remark:</u> If $D_1, \ldots, D_q$ are regular at W and at least one of the D_i is in $m\mathcal{D}_W$, then $\omega_q(D_1, \ldots, D_q) \in m$.

To see this we need only expand ω_q in terms of $d\xi_1, \ldots, d\xi_r$ and look at $\omega_q(D_1, \ldots, D_q)$. An immediate consequence of the remark is: <u>if</u> $D_1, \ldots, D_q \in \mathcal{D}^\circ_W$, <u>then</u> $\mathrm{Tr}_W \omega_q(D_1, \ldots, D_q)$ <u>depends only on the</u> $m\mathcal{D}_W$-<u>cosets of</u> $D_1, \ldots, D_q$, i.e., on the traces $\bar{D}_i = \mathrm{Tr}_W D_i$, $i = 1, \ldots, q$.

<u>Def. 10.5:</u> Let ω_q be a q-fold differential which is regular at W. If $q \leq s$, we define

$$\mathrm{Tr}_W \omega_q(\bar{D}_1, \ldots, \bar{D}_q) = \mathrm{Tr}_W \omega_q(D_1, \ldots, D_q)$$

where $\bar{D}_i = \mathrm{Tr}_W D_i$, $D_i \in \mathcal{D}^\circ_W$, $i = 1, \ldots, q$. If $q > s$, we fix q-s derivations $D'_1, \ldots, D'_{q-s}$ in $\mathcal{D}_W$ and we define $\mathrm{Tr}_W^{D'_1, \ldots, D'_{q-s}} \omega_q$ by

$$(\mathrm{Tr}_W^{D'_1, \ldots, D'_{q-s}} \omega_q)(\bar{D}_1, \ldots, \bar{D}_s) = \mathrm{Tr}_W(\omega_q(D_1, \ldots, D_s, D'_1, \ldots, D'_{q-s}))$$

where $D_i \in \mathcal{D}^\circ_W$, $\mathrm{Tr}_W D_i = \bar{D}_i$, $i = 1, \ldots, s$.

We now list some simple properties of the trace of a differential:

(1) $\mathrm{Tr}_W(\omega + \omega') = \mathrm{Tr}_W \omega + \mathrm{Tr}_W \omega'$

(2) $\mathrm{Tr}_W \omega \wedge \omega' = (\mathrm{Tr}_W \omega) \wedge (\mathrm{Tr}_W \omega')$.

(3) If $\xi \in \mathcal{O}_W(V/k)$, then $\mathrm{Tr}_W d\xi = d\bar{\xi}$ where $\bar{\xi} = \mathrm{Tr}_W \xi$.

Let $\{\xi_1, \ldots, \xi_r\}$ be a set of uniformizing coordinates of W, and let $\omega_q = \sum_{1 \leq j_1 < j_2 < \ldots < j_q \leq r} A_{(j)}\, d\xi_{j_1} \ldots d\xi_{j_q}$, $A_{(j)} \in \mathcal{O}$.

(4) If $q \leq s$, then $\mathrm{Tr}_W \omega_q = \sum \bar{A}_{(j)}\, d\bar{\xi}_{j_1} \ldots d\bar{\xi}_{j_q}$ where $\bar{A}_{(j)} = \mathrm{Tr}_W A_{(j)}$ and $\bar{\xi}_{j_i} = \mathrm{Tr}_W \xi_{j_i}$.

We turn now to the case $q > s$. It is clear that <u>if at least one of the derivations $D'_1, \ldots, D'_{q-s}$ is in $\mathcal{D}^\circ_W$, then</u> $\mathrm{Tr}_W^{D'_1, \ldots, D'_{q-s}} \omega_q = 0$. For, if $D'_1 \in \mathcal{D}^\circ_W$, then, since $\mathcal{D}^\circ_W / \mathfrak{m}\mathcal{D}_W$ is a free s-dimensional $\mathcal{O}/\mathfrak{m}$-module, we have $D_1, \ldots, D_s, D'_1$ linearly dependent (mod $\mathfrak{m}\mathcal{D}_W$) over $\mathcal{O}/\mathfrak{m}$. Therefore $\omega_q(D_1, \ldots, D_s, D'_1, \ldots, D'_{q-s}) \in \mathfrak{m}$ and so $\mathrm{Tr}_W^{D'_1, \ldots, D'_{q-s}} \omega_q = 0$.

Let $\{t_1, \ldots, t_{r-s}\}$ be a set of uniformizing parameters of W, and complete it to a set $\{\xi_1, \ldots, \xi_r\}$ of uniformizing coordinates of W where $\xi_{s+i} = t_i$, $i = 1, \ldots, r-s$. Then $\mathcal{D}_W = \sum_{\nu=1}^{r} \mathcal{O} \frac{\partial}{\partial \xi_\nu}$ and $\mathcal{D}^\circ_W = \sum_{\alpha=1}^{s} \mathcal{O} \frac{\partial}{\partial \xi_\alpha} + \mathfrak{m}\mathcal{D}_W$. Hence any trace of W will be linearly dependent (over $k(W)$) on

$$\mathrm{Tr}_W^{\partial/\partial \xi_{i_1+s}, \ldots, \partial/\partial \xi_{i_{q-s}+s}} \qquad (*)$$

where $1 < i_1 < i_2 < \ldots < i_{q-s} \leq r - s$. We write $(*)$ as

$$\mathrm{Tr}_W^{t_{i_1}, \ldots, t_{i_{q-s}}\,;\, \xi_1, \ldots, \xi_s}.$$

If $\omega_q = \sum A_{(j)} d\xi_{j_1} \ldots d\xi_{j_q}$, $1 \leq j_1 < j_2 < \ldots < j_q \leq r$, then

$$\mathrm{Tr}_W^{t_{i_j}; \xi_k} \omega_q = \bar{A}_{1, \ldots, s, i_1+s, \ldots, i_{q-s}+s}\, d\bar{\xi}_1 \ldots d\bar{\xi}_s .$$

Special case: q = r. If q = r, then $\omega = A d\xi_1 \dots d\xi_r$. Since q-s = r-s, we have $\mathrm{Tr}_W^{t_1,\dots,t_{r-s},\ \xi_1,\dots,\xi_s}\ \omega = \bar{A}\, d\bar{\xi}_1 \dots d\bar{\xi}_s$ where we note that $\bar{A} d\bar{\xi}_1 \dots d\bar{\xi}_s$ is not independent of the choice of $\xi_1, \dots, \xi_s$.

Special Case: q = r, s = r - 1, i.e. W is a prime cycle. In this case we have $\mathrm{Tr}_W^{t;\xi_1,\dots,\xi_{r-1}}\ \omega = \bar{A} d\bar{\xi}_1 \dots d\bar{\xi}_{r-1}$ and this is independent of the choice of the ξ_i. Thus if dim W = r-1, if ω is a differential of degree r which is regular on W and if t is a uni-fotmizing parameter of W, then $\mathrm{Tr}_W^t\, \omega$ is well-defined.

Let F be a non-singular surface, and let Γ be an irreducible curve on F. Let ω be a differential of degree two, and assume Γ is not a component of (ω). Finally, let t be a uniformizing parameter of Γ. Then $(\omega) + \Gamma - (t)$ is a cycle which does not have Γ as a component, and $\mathrm{Tr}_\Gamma[(\omega) + \Gamma - (t)]$ is a divisor in $k(\Gamma)$. Since $\mathrm{Tr}_\Gamma^t\, \omega \neq 0$, we can consider the divisor $(\mathrm{Tr}_\Gamma^t\, \omega)$. Let

$$s(\Gamma) = \mathrm{Tr}_\Gamma[(\omega) + \Gamma - (t)] - (\mathrm{Tr}_\Gamma^t\, \omega).$$

We shall show first that $s(\Gamma)$ <u>is independent of</u> ω <u>and</u> t. Let ω_1 be another differential satisfying the same conditions as ω. We can write $\omega_1 = A\omega$ where $A \in \mathfrak{o}$, $A \notin \mathfrak{m}$. Choose ξ so that $\{\xi, t\}$ is a set of uniformizing coordinates of Γ on F, and let $\omega = B d\xi\, dt$. Then $\mathrm{Tr}_\Gamma^t\omega = \bar{B} d\bar{\xi}$ and $\mathrm{Tr}_\Gamma^t\, \omega_1 = \overline{AB}\, d\bar{\xi}$. Hence

$$\mathrm{Tr}_\Gamma[(\omega_1) + \Gamma - (t)] - (\mathrm{Tr}_\Gamma^t\, \omega_1) = (\bar{A}) + \mathrm{Tr}_\Gamma[(\omega) + \Gamma - (t)] - (\bar{A}) - (\mathrm{Tr}_\Gamma^t\, \omega)$$
$$= s(\Gamma).$$

Let t_1 be another uniformizing parameter of W and let ξ be as before. We have $t_1 = Bt$ where B is a unit in $\mathfrak{o}$. Then $\left|\dfrac{\partial(t_1, \xi)}{\partial(t, \xi)}\right| = B + \dfrac{\partial B}{\partial t} t$

which is a unit in $\mathcal{O}$. Hence $\{\xi, t_1\}$ is also a set of uniformizing coordinates of W. Let $\omega = A d\xi dt = A_1 d\xi dt_1$; then $A = (B + \frac{\partial B}{\partial t} t)A_1$. Hence $\mathrm{Tr}_{\Gamma}^{t_1}\omega = \bar{A}_1 d\bar{\xi}$ and $\mathrm{Tr}_{\Gamma}^{t}\omega = \bar{A} d\bar{\xi} = \overline{(B + \frac{\partial B}{\partial t} t)A_1} d\bar{\xi} = \bar{B}\bar{A}_1 d\bar{\xi}$. It is clear now that $s(\Gamma)$ is independent of t as well as of ω.

Theorem 10.6: (a) The divisor $s(\Gamma)$ is effective, and its local component divisor at each point of Γ has even degree.

(b) The divisor $s(\Gamma) = 0$ if and only if Γ has no singular points. More precisely, a prime divisor $\bar{\gamma}$ of $k(\Gamma)$ is a component of $s(\Gamma)$ if and only if $\bar{\gamma}$ is centered at a singular point of Γ.

(c) Let $P \in \Gamma$; then $s(\Gamma)_P$ is the conductor of $\mathcal{O}_P(\Gamma/k)$ in its integral closure. (We shall not prove (c)).

Proof: Assume P is a simple point of Γ, and let $x = 0$ be a local equation of Γ at P (hence $x \in \mathfrak{m}_P(F/k)$). Consider $\mathcal{O}_P(F/k)/\mathcal{O}_P(F/k)x = \mathcal{O}_P(\Gamma/k)$. Since P is simple, $\mathcal{O}_P(\Gamma/k)$ is a regular local ring. Hence $\mathfrak{m}_P(\Gamma/k) = \mathfrak{m}_P(F/k)/\mathcal{O}_P(F/k)x$ is a principal ideal. Let $\mathfrak{m}_P(\Gamma/k) = (\bar{y})$ where $\bar{y}$ is the Γ-residue of some element $y \in \mathfrak{m}_P(F/k)$. Since $\{\bar{y}\}$ is a basis of $\mathfrak{m}_P(\Gamma/k)$, it follows that $\{x,y\}$ is a basis of $\mathfrak{m}_P(F/k)$. Therefore $\{x,y\}$ is a set of uniformizing parameters at P. This shows that we can always choose uniformizing parameters x,y at P such that $x = 0$ is a local equation of Γ at P.

Let $\{x,y\}$ be a set of uniformizing parameters at P where $x = 0$ is a local equation of Γ at P. Then $t = x$ is a uniformizing parameter of Γ on F. Since $s(\Gamma)$ is independent of ω, let $\omega = dxdy$, which is regular at Γ. Then $(\omega)_P = 0$ and $(t)_P = \Gamma$. Therefore $((\omega) + \Gamma - (t))_P = 0$, and so $\mathrm{Tr}_{\Gamma}((\omega) + \Gamma - (t))_P = 0$. Furthermore, since $\mathrm{Tr}_{\Gamma}^{x}\, dxdy = d\bar{y}$ and since $\bar{y}$ is a uniformizing parameter of P on

Γ, we have $(d\bar{y})_P = 0$. Thus $(\mathrm{Tr}^x_\Gamma \omega)_P = 0$. This shows that if Γ has no singularities, then $s(\Gamma) = 0$.

Assume P is a singular point of Γ. Let $\{x,y\}$ be a set of uniformizing parameters at P on F, and let $t = 0$ be a local equation of Γ at P. Again we can let ω = dxdy, and so we have as before $(t)_P = \Gamma$ and $(\omega)_P = 0$. Hence $\mathrm{Tr}_\Gamma((\omega) + \Gamma - (t))_P - (\mathrm{Tr}_\Gamma^{\ t}\,\omega)_P = -(\mathrm{Tr}^t_\Gamma\, dxdy)_P$. Let B be the set of prime divisors Υ of $k(\Gamma)$ which are centered at P. Let $\bar{\omega} = \mathrm{Tr}^t_\Gamma\, dxdy$; then $(\bar{\omega})_P = \sum_{\Upsilon \in B} v_\Upsilon(\bar{\omega}) \cdot \Upsilon$. To prove (a) we must show:

(1) $v_\Upsilon(\bar{\omega}) < 0$ for all $\Upsilon \in B$, and

(2) $\sum_{\Upsilon \in B} v_\Upsilon(\bar{\omega})$ is even.

We adopt the following notations: $\mathcal{O} = \mathcal{O}_P(F/k)$, $M = \mathfrak{m}_P(F/k)$, $\mathcal{O}' = \mathcal{O}_\Gamma(F/k)$ and $\mathfrak{m} = \mathfrak{m}_\Gamma(F/k)$.

We can consider M/M^2 as a vector space over k. For $\xi \in M$, there exist unique elements c,d such that $\xi - (cx + dy) \in M^2$. Thus we can look on M/M^2 as the two-dimensional set of linear forms $cX + dY$. The elements of the corresponding one-dimensional projective space are definable by equations of the form $cX + dY = 0$. We call the elements of this projective line <u>directions at P</u>.

Let $\xi \in M$; there exists an integer n, $n \geq 1$, such that $\xi \in M^n$ and $\xi \notin M^{n+1}$. Hence we can write $\xi = a_0x^n + a_1x^{n-1}y + \ldots + a_ny^n$ where $a_i \in \mathcal{O}$, $i = 0, \ldots, n$, and not all the a_i are in M. Let $\bar{a}_i$ denote the M-residue of a_i (hence $\bar{a}_i \in k$). We call $\bar{a}_oX^n + \bar{a}_1X^{n-1}Y + \ldots + \bar{a}_nY^n$ the <u>leading form of</u> ξ. A well-known characteristic property of uniformizing parameters of a regular local ring is: if $b_ox^n + b_1x^{n-1}y + \ldots + b_ny^n = 0$, then $b_i \in M$ fpr all i. Hence we see that the leading form of ξ is independent of the representation $\xi = a_ox^n + a_1x^{n-1}y + \ldots + a_ny^n$

If $cX + dY$ is a factor of the leading form of ξ, then the direction $cX + dY$ is said to be <u>tangent to the cycle</u> $(\xi)_P$

Let $\bar{x}$ and $\bar{y}$ by the Γ-traces of x and y, and let $\gamma \in B$. Since γ is centered at P, we have $v_\gamma(\bar{x}) > 0$ and $v_\gamma(\bar{y}) > 0$. Furthermore, since we can always replace x and y by the uniformizing parameters x and $y - cx$ where c is a unit in $\mathcal{O}$ (i.e., change x and y by a non-singular linear transformation), we can assume $v_\gamma(\bar{x}) < v_\gamma(\bar{y})$. Then we assert that $Y = 0$ <u>is tangent to</u> Γ. For we have $t \in M^s$, $t \notin M^{s+1}$ for some integer s. Therefore $t - (c_0x^s + c_1x^{s-1}y + \dots + c_sy^s) \in M^{s+1}$ where $c_0X^s + c_1X^{s-1}Y + \dots + c_sY^s$ is the leading form of t. We can write $t = a_0x^s + a_1x^{s-1}y + \dots + a_sy^s$ where $a_i \in \mathcal{O}$, $i = 0, \dots, n$. Then Y will be a factor of the leading form of t if $a_0 \in M$. Taking residues mod $\mathcal{O}t$, we have $\bar{a}_0\bar{x}^s + \bar{a}_1\bar{x}^{s-1}\bar{y} + \dots + \bar{a}_s\bar{y}^s = 0$. If $\bar{a}_0$ is a unit in $\mathcal{O}_P(\Gamma/k) = \mathcal{O}/\mathcal{O}t$, then $v_\gamma(\bar{a}_0) = 0$. But this yields an obvious contradiction with the fact that $v_\gamma(\bar{x}) < v_\gamma(\bar{y})$. Hence $a_0 \in M$ and $Y = 0$ is tangent to Γ.

Let $\mathcal{O}_1 = \left\{\frac{\xi}{\eta} \middle| \xi, \eta \in \mathcal{O}; \eta \in M^n, \xi \notin M^{n+1}, \xi \in M^n\right.$ and $Y = 0$ not tangent to $\left.(\eta)\right\}$. A well-known result for regular local rings is: if $\alpha, \beta \in \mathcal{O}$ where $\alpha \in M^h$, $\alpha \notin M^{h+1}$, $\beta \in M^g$, $\beta \notin M^{g+1}$, then $\alpha\beta \notin M^{g+h+1}$. Hence $\mathcal{O}_1$ is a ring. Clearly $\mathcal{O} \subset \mathcal{O}_1$. We shall now show that, in fact, $\mathcal{O}_1$ <u>is a regular local ring of dimension two.</u>

Let $\zeta = \xi/\eta \in \mathcal{O}_1$. Then ζ is a non-unit if and only if

(a) $\xi \in M^{n+1}$; or

(b) $\xi \notin M^{n+1}$, and $Y = 0$ is tangent to (ξ).

It follows that the non-units form an ideal M_1 in $\mathcal{O}_1$.

It is clear that $x \in M_1$. Let $y_1 = y/x$; then y_1 is not only in $\mathcal{O}_1$ but even in M_1. We claim that x and y_1 form a basis of M_1.

Let $\zeta = \xi/\eta \in M_1$. We can write $\xi = b_0x^n + b_1x^{n-1}y + \dots + b_ny^n$ where $b_0 \in M$. Since $M = \mathcal{O}(x,y)$, we see that $\mathcal{O}_1 M = \mathcal{O}_1 x$. Hence $\xi \in x^n\mathcal{O}_1(x,y_1)$, and so we have $\xi = x^n\xi_1$ where $\xi_1 \in \mathcal{O}_1(x,y_1)$. We can write $\eta = (\frac{\eta}{x^n})x^n$. Since $\eta \in M^n$ and $Y = 0$ is not tangent to (η), we have $\eta/x^n \in \mathcal{O}_1$. Therefore $\eta = \eta_1x^n$ is a unit in $\mathcal{O}_1$. Hence $\xi/\eta \in \mathcal{O}_1(x,y_1)$ which shows that x and y_1 generate M_1.

Consider $\mathcal{O}[y_1] \subset \mathcal{O}_1$, and let $H = \{a_0 + a_1y + \dots a_ny^n \mid a_i \in \mathcal{O}, a_0 \notin M\}$. It is clear that H is a multiplicatively closed set and that $\mathcal{O}_1 = \mathcal{O}[y_1]_H$. Since $\mathcal{O}[y_1]$ is noetherian, it follows that $\mathcal{O}_1$ is noetherian and hence $\mathcal{O}_1$ is a local ring. To prove $\mathcal{O}_1$ is regular we must show $\dim_k M_1/M_1^2 = 2$.

It is clear that $\dim_k M_1/M_1^2 \leq 2$ since x and y_1 generate M_1. To show that $\dim_k M_1/M_1^2 = 2$, it suffices to prove the following assertion: if $Ax + By_1 = 0$ where $A,B \in \mathcal{O}_1$, then $A,B \in M_1$. We can write $A = \xi/\eta$, $B = \xi_1/\eta$ where $\xi, \xi_1, \eta \in M^n$; $\eta \notin M^{n+1}$ and $Y = 0$ is not tangent to (η). Then $\xi x^2 + \xi_1 y = 0$. Since $\xi x^2 \in M^{n+2}$, $y \in M$ and $y \notin M^2$, we see that $\xi_1 \in M^{n+1}$. Therefore B is a non-unit in $\mathcal{O}_1$, i.e., $B \in M_1$. Furthermore, since $Y = 0$ is tangent to (ξ), we see that $\xi/\eta = A$ is a non-unit in $\mathcal{O}_1$. This shows $\dim_k M_1/M_1^2 = 2$, and therefore $\mathcal{O}_1$ is a regular local ring.

Let $\mathfrak{p} = \mathcal{O}t$; then $\sigma = \mathcal{O}_{\mathfrak{p}}$ and hence $\mathcal{O} \subset \sigma$. Let $\xi/\eta \in \mathcal{O}_1$; then $\eta \in M^n$, $\eta \notin M^{n+1}$ and $Y = 0$ is not tangent to (η). This means $\eta \not\equiv 0 \pmod t$ since $Y = 0$ is tangent to t. Hence $\xi/\eta \in \sigma$ and $\mathcal{O}_1 \subset \sigma$.

Let $\mathfrak{p}_1 = \mathfrak{m} \cap \mathcal{O}_1$ where we recall that $\mathfrak{m} = \mathfrak{m}_{\Gamma}(F/k)$. Then $\mathfrak{p}_1$ is a prime ideal in $\mathcal{O}_1$, and $\mathfrak{m} \cap \mathcal{O} = \mathfrak{p} = \mathcal{O}t$. Hence $\mathfrak{p}_1 \cap \mathcal{O} = \mathfrak{p}$. This yields the following sequence of inclusions $\sigma = \mathcal{O}_{\mathfrak{p}} \subset \mathcal{O}_{1\mathfrak{p}_1} \subset \sigma_{\mathfrak{m}} = \sigma$. Therefore $\sigma = \mathcal{O}_{1\mathfrak{p}_1}$.

We have defined $\mathfrak{p} = \mathcal{O}t$. There is an integer s such that $t \in M^s$ and $t \notin M^{s+1}$. Viewing t as in M_1, we can write $t = x_1{}^s t_1$ where $t_1 \in \mathcal{O}_1$ and $\{x_1, y_1\}$ is a set of uniformizing parameters of M_1. Since $x_1{}^s t_1 \in \mathfrak{p}_1$ and $x_1 \notin \mathfrak{p}_1$, we have $t_1 \in \mathfrak{p}_1$. We shall show that $\mathfrak{p}_1 = \mathcal{O}_1 t_1$. Clearly $\mathcal{O}_1 t_1 \subset \mathfrak{p}_1$. Let $\zeta = \xi/\eta \in \mathfrak{p}_1$; then $\zeta \in \mathfrak{m}$. Since $\eta \notin \mathfrak{m}$, we see that $\xi \in \mathcal{O}t$. $\eta_1 = \eta/x_1{}^n$ is a unit in $\mathcal{O}_1$, and $\xi = \xi' t$ where $\xi' \in M^{n-s}$. Therefore $\xi' = x_1{}^{n-s}\xi_1$ where $\xi_1 \in \mathcal{O}_1$, and $\xi = x_1{}^n \xi_1 t_1$. Hence $\zeta = (\xi_1/\eta_1)t_1 \in \mathcal{O}_1 t_1$, and so $\mathcal{O}_1 t_1 = \mathfrak{p}_1$. This shows that $t_1 = 0$ is a local equation of Γ in $\mathcal{O}_1$.

What we have described above is the effect, in a given direction at $\mathcal{O}$, of a so-called "locally quadratic transformation" T of F with center P. A full (global) description of T is as follows:

Let $y_0, \ldots, y_n$ be (strictly) homogeneous coordinates of the general point A of F/k, and let us assume that the given point P is the point $Y_0 = 1, Y_1 = \ldots = Y_n = 0$. Consider, in projective space of dimension $(\frac{1}{2})(n+1)(n+2) - 2$, the point A' whose homogeneous coordinates are the products $y_i y_j$, $0 \leq i \leq j \leq n$, $(i,j) \neq (0,0)$, and let F' be the locus of A' over k. Then F' is a surface, and the algebraic correspondence with general point (A, A') is a birational transformation $T : F \to F'$ of F onto F'. The following properties of T are easily established:

(1) The total transform $T\{P\}$ of P is an irreducible, non-singular rational curve Δ' on F'.

(2) T is biregular on $F - P$ and T^{-1} is regular on F' (biregular on $F' - \Delta'$).

(3) There is a one-to-one correspondence between the directions of F at P and the points of Δ'.

If $L = aX + bY = 0$ is a direction at P and if $\mathcal{O}'_{P'}$ is the local ring $\mathcal{O}_{P'}(F'/k)$ of the corresponding point P' of $F'(P' \in \Delta')$, then $\mathcal{O}'_{P'} = \{\xi/\eta \mid \xi,\eta \in M^n,\ \eta \notin M^{n+1}$ and L is not a tangent to $(\eta)_P\}$. From the preceding "local analysis" it follows that F' is a non-singular surface (in particular, $\mathcal{O}'_{P'}$ is a regular local ring for each point P' of Δ'), and that to the curve Γ there will correspond on F' an irreducible curve Γ' whose intersections with Δ' are the points $P'_1, \ldots, P'_h$ which correspond to the tangent directions of Γ at P.

We recall that B is the set of valuations of $K(\Gamma)$ with center P. Let $B = B_1 \sqcup B_2 \sqcup \ldots \sqcup B_h$ where the B_i contain only valuations with the same tangential direction.

We recall that $\{x,y\}$ is a set of uniformizing parameters of M. Let P'_1 be one of the h points P'_i, and let $\{x_1,y_1\}$ be a set of uniformizing parameters of M_1 where we may assume that $x_1 = x$ and $y = x_1y_1$. Then $dx\,dy = x_1dx_1dy_1$. Therefore

$$\mathrm{Tr}_{\Gamma}^{t}\,dxdy = \mathrm{Tr}_{\Gamma}^{x_1^s t_1}\,dxdy = (1/\overline{x}_1^{\,s})\,\mathrm{Tr}_{\Gamma}^{t_1}\,dxdy$$

where $\overline{x} = \mathrm{Tr}_{\Gamma}\,x$. Hence $\mathrm{Tr}_{\Gamma}^{t}\,dxdy = (1/\overline{x}_1^{\,s-1})\mathrm{Tr}_{\Gamma}^{t_1}\,dx_1dy_1$.

Let $\gamma \in B_1$. Since we can assume $v_\gamma(\overline{y}) > v_\gamma(\overline{x})$, it follows that $v_\gamma(\overline{x}_1) > 0$ and $v_\gamma(\overline{y}_1) > 0$. Therefore

$$v_\gamma(\mathrm{Tr}_{\Gamma}^{t}\,dxdy) = -(s-1)v_\gamma(\overline{x}) + v_\gamma(\mathrm{Tr}_{\Gamma}^{t_1}\,dx_1dy_1). \qquad (*)$$

To each B_i there corresponds a local ring $\mathcal{O}_i$ (of the point P'_i) with uniformizing parameters x_i, y_i. For each $i = 1,2, \ldots, h$ we obtain an expression analogous to (*) and $\overline{x}$ is the same in each case (for we may assume that $v(\overline{y}) \geq v(\overline{x})$, for all $v \in B$). Therefore

$$(\mathrm{Tr}_{\Gamma}^{t}\,dxdy) = -(s-1)(\overline{x})_P + \sum_{i=1}^{h}(\mathrm{Tr}_{\Gamma'}^{t_i}\,dx_idy_i)_{P'_i}\ .$$

We can assume that $\{t,x\}$ is a set of uniformizing coordinates of Γ on F. Then $\frac{\partial(x,t)}{\partial(y,t)} \notin \mathfrak{m}$, and therefore either $\frac{\partial t}{\partial y} \notin \mathfrak{m}$ or $\frac{\partial t}{\partial x} \notin \mathfrak{m}$. Hence $Tr_\Gamma \frac{\partial t}{\partial x}$ and $Tr_\Gamma \frac{\partial t}{\partial y}$ are not both zero. Let

$$c(\Upsilon;P) = \min\left\{v_\Upsilon(Tr_\Gamma \frac{\partial t}{\partial x}),\ v_\Upsilon(Tr_\Gamma \frac{\partial t}{\partial y})\right\} .$$

Then $c(\Upsilon;P) = 0$ a non-negative integer since $\frac{\partial t}{\partial x}, \frac{\partial t}{\partial y} \in \mathcal{O}$. Hence $c(\Upsilon;P) = 0$ if and only if $s = 1$ and P is therefore a simple point. We shall finish the proof of the theorem by induction on $c(\Upsilon;P)$.

Assume $s > 1$. We know $c(\Upsilon;P) \geq (s-1)v_\Upsilon(\bar{x}) + c(\Upsilon;P_1')$. Therefore $c(\Upsilon;P) > c(\Upsilon;P_1')$ for $s > 1$. Thus by the induction hypothesis $v_\Upsilon(Tr_\Gamma^{t_1} dx_1dy_1) < 0$ and so $v_\Upsilon(Tr_{\Gamma'}^{t} dxdy) < 0$.

The induction hypothesis also yields the fact that $\Sigma(Tr_\Gamma^{t_1} dx_1dy_1)_P$ has even degree. By an earlier result we know that $\deg(\bar{x})_P = i(\Gamma,\Delta;P) = \dim_k \mathcal{O}/\mathcal{O}(t,x)$ where Δ is $x = 0$. Now we can write $t = a_0x^s+a_1x^{s-1}y + \dots + a_sy^s$ where we must have $a_s \notin M$. Therefore the set $\{1,y,y^2, \dots, y^{s-}$ is a basis of $\mathcal{O}/\mathcal{O}(t,x)$ over k. This shows that $\dim_k \mathcal{O}/\mathcal{O}(t,x) = s$.

Hence

$$\deg(Tr_\Gamma^t dxdy)_P = -(s-1)s + \text{ an even integer},$$

and so $\deg(Tr_\Gamma^t dxdy)_P$ is an even integer.

§11. The arithemetic genus

Let F be a non-singular surface and Γ an irreducible algebraic curve on F. Let $K = (\omega)$ be any canonical divisor, and let Z be a divisor which is linearly equivalent to $K + \Gamma$ and does not have Γ as a component. For example, $Z = \Gamma + (\omega) - (t)$ where t is a uniformizing parameter of Γ. Then $Tr_\Gamma Z = \overline{K} + X$ where X is an effective divisor and $\overline{K}$ is a canonical divisor on Γ. If Γ has no singular points, then $Tr_\Gamma Z = \overline{K}$.

We restate Theorem 10.6 and the paragraph above as

<u>Prop. 11.1</u>: If Γ is an irreducible curve on F, K a canonical divisor on F and Z a cycle such that

(1) $Z \equiv \Gamma + K$ and

(2) Γ is not a component of Z

then the divisor $Tr_\Gamma Z$ of $k(\Gamma)$ has the form

$$Tr_\Gamma Z = K(\Gamma) + S(\Gamma)$$

where $K(\Gamma)$ is a canonical divisor of $k(\Gamma)$ and $s(\Gamma)$ (the "divisor of the singularities" of Γ) is an effective cycle each local P-component of which has even degree. A prime divisor γ of $k(\Gamma)$ is a component of $S(\Gamma)$ if and only if γ is centered at a singular point of Γ. In particular, $S(\Gamma) = 0$ if and only if Γ has no singular points.

Let $\pi(\Gamma)$ be the (effective) genus of Γ. Then $(Z.\Gamma) - 2\pi(\Gamma) + 2$ is a non-negative even integer. Furthermore, if Γ is an irreducible curve on F and K is a canonical divisor on F, then $(K.\Gamma) + (\Gamma^2) - 2\pi(\Gamma) + 2$ is a non-negative even integer. We can therefore define $p(\Gamma)$ by the equation

$$(K.\Gamma) + (\Gamma^2) = 2p(\Gamma) - 2.$$

Then (1) $p(\Gamma)$ is an integer,

(2) $p(\Gamma) \geq \pi(\Gamma)$ and

(3) $p(\Gamma) = \pi(\Gamma)$ if and only if Γ has no singularities.

We are thus led to the following

<u>Def. 11.2:</u> If Z is any cycle on F, then the <u>arithmetic genus</u> $p(z)$ of Z is defined by $(K.Z) + (Z^2) = 2p(Z) - 2$.

<u>Note:</u> (a) $p(0) = 1$

(b) $p(K) = (K^2) + 1$

(c) $p(-K) = 1$.

<u>Prop. 11.3:</u> (a) $p(Z_1+Z_2) = p(Z_1) + p(Z_2) + (Z_1.Z_2) - 1$.

(b) $p(Z)$ is an integer.

(c) If Z is an irreducible curve, i.e., a prime cycle, then $p(Z) \geq \pi(Z)$. Furthermore $p(Z) = \pi(Z)$ if and only if Z has no singular points.

Proof: (a) $2p(Z_1+Z_2) = (K.Z_1)+(K.Z_2)+(Z_1^2)+2(Z_1.Z_2)+(Z_2^2)+2$. Hence $p(Z_1+Z_2) = p(Z_1) + p(Z_2) + (Z_1.Z_2) - 1$.

(c) has already been proven.

(b) By virtue of (a) we can assume Z is a curve or the negative of a curve. If Z is a curve we have already shown that p(Z) is an integer. Hence assume $Z = -\Gamma$ where Γ is a curve. Since p(O) = 1, we have, using (a),

$$p(\Gamma + (-\Gamma)) = 1 = p(\Gamma) + p(-\Gamma) - (\Gamma^2) - 1.$$

Hence $p(-\Gamma)$ is an integer.

Assume Γ is a non-singular curve and consider $|\Gamma + K|$ (the "adjoint system of $|\Gamma|$"). We have

$$\dim |\Gamma + K| = \dim |K| + \dim Tr_{\Gamma} |\Gamma + K| + 1.$$

By definition $Tr |\Gamma + K| = \{ Tr_{\Gamma} Y | Y \in |\Gamma + K|$, Γ is not a component of $Y \}$, and we have shown that if Γ is non-singular, then $Tr_{\Gamma} Y$ is a canonical divisor. Hence $Tr_{\Gamma} |\Gamma + K|$ is a subsystem of the canonical system on Γ. If Γ is non-singular, the dimension of the canonical system on Γ is $p(\Gamma) - 1$. Therefore $-1 \leq Tr_{\Gamma} |\Gamma + K| \leq p(\Gamma) - 1$. Let $\dim Tr_{\Gamma} |\Gamma + K| = p(\Gamma) - 1 - \delta(\Gamma)$ where $0 \leq \delta(\Gamma) \leq p(\Gamma)$. Then since $\dim |K| = p_g - 1$, we have

$$\dim |\Gamma + K| = p_g + p(\Gamma) - 1 - \delta(\Gamma).$$

§12. Normalization and complete systems.

Let $V \subset S_n$, and let $(y_o, \ldots, y_n)$ be strictly homogeneous coordinates of a general point of V. Let $R = k[V] = k[y_o, \ldots, y_n]$; then R is a graded ring; $R = R_o + R_1 + R_2 + \ldots + R_q + \ldots$ where R_q is the set of homogeneous polynomial expressions of degree q. Finally, let $K = k(y)$ (note: K is not $k(V)$).

Def. 12.1: An element $\xi \in K$ is homogeneous if $\xi = f_q(y)/f_s(y)$ where $f_q(y)$ and $f_s(y)$ are forms of degree q and s respectively. The degree of ξ is $q-s$ (and depends only on ξ).

Let K_q be the set of homogeneous elements of K of degree q (hence $K_o = k(V)$). Without loss of generality we can assume $y_o \neq 0$. If $\xi \in K_q$, then $\xi/y_o^q \in K_o$. Thus we see that $K_q = K_o y_o^q$ and $\Sigma K_q = K_o[y_o, 1/y_o]$. Since y_o is transcendental over K_o, ΣK_q is a direct sum and hence a graded ring.

Let R' be the integral closure of R in K. Since $R \subset K_o[y_o]$ and y_o is transcendental over K_o, it follows that $R' \subset K_o[y_o]$. Hence R' is a graded ring and we can write $R' = \sum_{q \geq 0} R'_q$ where $R'_q = R' \cap K_q$ and $\Sigma R'_q$ is a direct sum. R' is a finitely-generated R-module, so $R' = Rz_1 + \ldots + Rz_h$ where we can assume the z_i are homogeneous. Hence R'_q is a finite-dimensional vector space over k for each q. Let $R'_q = ku_o + \ldots + ku_m$ and consider $(u_o, \ldots, u_m)$ as the coordinates of a point U in S_m. It is clear that they are strictly homogeneous coordinates of U. Let V' be the locus of U in S_m.

Let A/k and A'/k be birationally equivalent varieties, and let $L = k(A) = k(A')$. Let B and B' be irreducible subvarieties of A and A' respectively. We say B and B' are corresponding varieties if there exists a valuation v of L/k such that B is the center of v on A and

B' is the center of v on A'. We say the birational transformation $A \to B$ has no <u>fundamental varieties</u> if to any irreducible subvariety of A there corresponds only a finite number of irreducible subvarieties of A'.

<u>Prop. 12.2:</u> The varieties V and V' are birationally equivalent, the birational transformation $V' \to V$ is regular and the birational transformation $V \to V'$ has no fundamental varieties.

Proof: Since the u_i are homogeneous elements of the same degree, $u_i/u_j \in K_0$ for all i and all j. Hence $k(U) \subset K_0 = k(V)$. It is clear from considering the elements $y_0^q, y_0^{q-1}y_1, \ldots, y_0^{q-1}y_n$ (which are linear combinations of the u_j) that $y_i/y_0 \in k(U)$, $i = 0, \ldots, n$, and therefore $k(V) = k(V')$. Thus V and V' are birationally equivalent.

Let $V_i = V - (V \cap H_i)$ where H_i is given by $Y_i = 0$, $i = 0, \ldots, n$. We have $y_i^q = \sum_{j=0}^{m} c_{ij}u_j$, $i = 0, \ldots, n$. Let $V_i' = V' - (V' \cap H_i')$ where H_i' is given by $\sum_{j=0}^{m} c_{ij}Y_j' = 0$, $i = 0, \ldots, n$. We shall now show that

(a) $k[V_i']$ is integrally dependent on $k[V_i]$; and

(b) $V_0', \ldots, V_n'$ cover V'.

We have $k[V_0] = k[y_1/y_0, \ldots, y_n/y_0]$ and $k[V_0'] = k[u_0/y_0^q, \ldots, u_m/y_0^q]$. To show (a) we must show u_i/y_0^q is integral over $k[V_0]$. Since u_i is in R', we can write $u_i^\nu + a_1(y)u_i^{\nu-1} + \ldots + a_\nu(y) = 0$, where $a_i(y) \in R$. Since u_i is homogeneous of degree q, we can assume all the terms are of the same degree, i.e. $a_j(y)$ is homogeneous of degree jq. Then we have

$$\left(\frac{u_i}{y_0^q}\right)^{\nu} + \frac{a_1(y)}{y_0^q}\left(\frac{u_i}{y_0^q}\right)^{\nu-1} + \ldots + \frac{a_\nu(y)}{y_0^{\nu q}} = 0.$$

This shows u_i/y_0^q is integral over $k[V_0]$ and proves (a).

It is sufficient to prove (b) for a rational point P' of V'. There is a zero-dimensional valuation v with center P' on V'. Let P be the center of v on V. Since the V_i cover V, we have $P \in V_i$ for some i. Hence $R_v \supset \mathcal{O}_P(V_i/k) \supset k[V_i]$. Now R_v is integrally closed, and thus by (a) we have $k[V_i'] \subseteq R_v$. Hence $P' \in V_i'$ and therefore the V_i' cover V'.

It is clear that $k[V_i'] \supset k[V_i]$; hence the birational transformation $V' \to V$ is regular.

To finish the proof we shall need the following well-known result:

(1) If A and A' are two noetherian rings such that $A \subseteq A'$ and A' is integral over A, then given any prime ideal $\mathfrak{p}$ in A there are only a finite number of prime ideals in A' which contract to $\mathfrak{p}$ in A.

Let $W \subset V$ and assume V_0 carries an affine representative of W. Let W' be a corresponding variety on V'. Then there exists a valuation v of $k(V) = k(V')$ with the centers W and W' on V and V' respectively. We have $R_v \supset \mathcal{O}_W(V/k) \supset k[V_0]$. Hence $R_v \supset k[V_0']$ and so V_0' carries an affine representative of W'.

Let $A = k[V_0]$, let $A' = k[V_0']$, let $\mathfrak{p}$ be the prime ideal of W in A and let $\mathfrak{p}'$ be the prime ideal of W' in A'. Since both W and W' are centers of v, we have $\mathfrak{p}' \cap A = \mathfrak{p}$. Applying (1) we have proved the proposition.

<u>Prop. 12.3</u>: Let W be an irreducible subvariety of V; and let $W_1', \ldots, W_g'$ be the irreducible subvarieties of V' which correspond to W. Then $\dim W = \dim W_i'$, $i = 1, \ldots, g$; and if W has an affine representative on V_0, then $\bigcap_{i=1}^{g} \mathcal{O}_{W_i'}(V'/k) = A_{\mathfrak{p}}'$ where $A_{\mathfrak{p}} = \mathcal{O}_W(V/k)$, and $A_{\mathfrak{p}}'$ is integral over $A_{\mathfrak{p}}$.

Cor. 12.4: The group of divisors of V coincides with the group of divisors of V'^q $(q \geq 1)$.

Theorem 12.5: If q is large, then V'^q is arithmetically normal.

Proof: Let $R = k[y_0, \ldots, y_n]$, and let R' be the integral closure of R. Then we can write $R' = R_0 + R_1 + \ldots + R_q + \ldots$ and $R'_q = R'_0 + R'_1 + \ldots + R'_q + \ldots$. Let $I = k[u_0, \ldots, u_m]$ where $R' = \sum_{i=0}^{m} ku_i$. (Note: $I \subseteq R'$.) We must show that I is integrally closed in its quotient field for large q.

Since I is a graded ring, we can write $I = I_0 + I_1 + \ldots + I_h + \ldots$ noting that $I_h \subset R'_{hq}$. Let I' be the integral closure of I in its quotient field, then $I' = I'_0 + I'_1 + \ldots + I'_h + \ldots$ where $I'_h \subset R'_{hq}$. We assert that $I'_h = R'_{hq}$. The elements of R'_{hq} are integral over R and therefore over I. To show $I'_h = R'_{hq}$, we must show that R'_{hq} is contained in the quotient field of I. Let ξ be in R'_{hq}.

Then

$$\xi = \frac{C_{(p+1)hq}(y)}{D_{pqh}(y)} = \frac{C'_{(p+1)h}(u)}{D'_{ph}(u)},$$

and therefore ξ is in the quotient field of I. Hence we have to show $I' = R'_0 + R'_q + R'_{2q} + \ldots + R'_{hq} + \ldots$.

Since R' is a finitely-generated R-module, we can write $R' = Rz_1 + \ldots + Rz_t$ where we can assume the z_i are homogeneous and of degree s_i. Let $s = \max(s_1, \ldots, s_t)$. We shall show that if $q \geq s$, then V'^q is arithmetically normal, i.e., $I = I'$.

For $q \geq s$ we clearly have $R'_q = R_{q-s_1}z_1 + \ldots + R_{q-s_t}z_t$. Let j be any non-negative integer. Then

$$\begin{aligned} R'_{q+j} &= R_{q+j-s_1}z_1 + \ldots + R_{q+j-s_t}z_t \\ &= R_jR_{q-s_1}z_1 + \ldots + R_jR_{q-s_t}z_t \\ &= R_jR'_q . \end{aligned}$$

It follows that we also have $R'_{q+j} = R'_q R'_j$. In particular, $R'_{qh} = (R'_q)^h$, and therefore $I' = k + R'_q + (R'_q)^2 + \dots + (R'_q)^h + \dots$. Recalling that $I = k[u_0, \dots, u_m]$ where $R'_q = \sum_{i=0}^{m} ku_i$, we see that $I' \subseteq I$ and hence $I' = I$. This proves the theorem.

The process of passing from V to V'^q is called a <u>normalization</u> of V, and V'^q is called a <u>derived normal model</u>.

Let S be a finite k-module of homogeneous elements of K of degree q, i.e., $S \subset K_q = K_0 y_0{}^q$ where $K_0 = k(V)$. Fix $\xi_0 \in S$, $\xi_0 \neq 0$, and let $M = \{\xi/\xi_0 / \xi \in S\} \subseteq K_0$. We know there exists a smallest effective cycle Δ such that $(\xi/\xi_0) + \Delta \geq 0$. The set $\{(\xi/\xi_0) + \Delta \mid \xi \in S\}$ is called the linear system of M and is denoted by $LS(M)$. (See section 5.
Let Z_ξ be defined by $(\xi/\xi_0) = Z_\xi - \Delta$. Setting $\xi = \xi_0$, we see that $\Delta = Z_{\xi_0}$. Hence we can write $(\xi/\xi_0) = Z_\xi - Z_{\xi_0}$. Similarly we get $(\xi/\xi_1) = Z_\xi - Z_{\xi_1}$. Thus we have a linear system $L(S) = \{Z_\xi \mid Z_\xi - Z_{\xi'} = (\xi/\xi')\}$. The system $L(S)$ has no fixed components and is uniquely determined by this condition and by the given k-module S.

If $S = R_q$, i.e., if S is the set of forms $f(y_0, \dots, y_n)$ of degree q, then $L(S)$ is called <u>the linear system cut out on</u> V <u>by the hypersurfaces of order</u> q. If $f(y) \neq 0$, then we have a cycle Z_f called the <u>intersection cycle of</u> V <u>with the hypersurface</u> $f(Y) = 0$. We write L_q for $L(R_q)$ and L'_q for $L(R'_q)$.

<u>Prop. 12.6</u>: Let S and S' be two finite k-modules of homogeneous functions of degree q where $S \subset S'$. If the elements of S' are integral over $k[S]$, then $L(S) \subset L(S')$.

Proof: Fix $\xi_0 \in S$, $\xi_0 \neq 0$. For $\xi \in S$ and $\xi' \in S'$ we have cycles Z_ξ and $Z'_{\xi'}$ such that

$$Z_\xi - Z_{\xi_0} = (\xi/\xi_0) \quad\text{and}\quad Z'_{\xi'} - Z'_{\xi_0} = (\xi'/\xi_0).$$

Since Z_{ξ_0} is the smallest effective cycle Δ with the property that $(\xi/\xi_0) + \Delta \geq 0$ and since $S \subset S'$, it follows that $Z_{\xi_0} \leq Z'_{\xi_0}$. Hence, in order to prove the proposition, we must show only that $(\xi'/\xi_0) + Z_{\xi_0} \geq 0$, for all ξ' in S'. Let $S = \Sigma k v_j$. Since ξ' is homogeneous and integral over $k[S]$, we can write

$$\xi'^{\nu} + a_1(v)\xi'^{\nu-1} + \dots + a_{\nu}(v) = 0 \qquad (1)$$

where $a_j(v)$ is homogeneous of degree j and $a_j(v) \in k[S]$. Since v_o is in S, we can choose $\xi_o = v_o$. Dividing (1) by v_o^{ν}, we see that $\eta' = \xi'/v_o$ is integral over $k[v_1/v_o, \dots, v_m/v_o]$. Hence there exist polynomials $A_j(\frac{v_i}{v_o})$, where the degree of A_j is $\leq j$, such that

$$\eta'^{\nu} + A_1(v/v_o)\eta'^{\nu-1} + \dots + A_{\nu}(v/v_o) = 0 \qquad (2)$$

Let Γ occur in Z_{v_o} with coefficient μ $(\mu \geq 0)$. Then $v_{\Gamma}(v_j/v_o) \geq -\mu$ for all j and so $v_{\Gamma}(A_j(\frac{v}{v_o})) \geq -j\mu$. Therefore, by (2) we see that $v_{\Gamma}(\eta') \geq -\mu$ and this shows $(\xi'/v_o) + Z_{v_o} \geq 0$ which proves the proposition.

<u>Theorem 12.8</u>: $L'_q = |L_q|$ where $|L_q|$ denotes the complete linear system determined by L_q.

Proof: Since R'_q is integral over R_q, Prop. 12.6 shows that $L_q \subseteq L'_q$.

Let $Z(f)$ denote Z_f. Then $Z(y_i{}^q) = q\,Z(y_i) = q\,C_i$ where $C_i = Z(y_i)$, $i = 0, \dots, n$. Let D be any cycle in $|L_q|$; then $D \equiv qC_i$ for $i = 0, \dots, n$. Let $D - qC_i = (\xi_i)$, $\xi_i \in k(V)$. Then $(\xi_i/\xi_j) = (y_j{}^q/y_i{}^q)$, and for a suitable choice of the ξ_i, we have $\xi_i/\xi_j = y_j^q/y_i{}^q$. Therefore $\xi_i y_i{}^q = \xi_j y_j{}^q$ for all i,j. By definition $(\xi_o) = D - qC_o$, hence <u>the poles of</u> ξ_o <u>are among the components of</u> C_o.

Consider $S = k[y_1/y_o, \dots, y_n/y_o]$, and let $\overline{S}$ be the integral closure of S. Let $\mathfrak{R}$ be the set of all divisors v such that $S \subset R_v$ and $M_v \cap S$ is a minimal prime ideal in S. Then $\overline{S} = \bigcap_{v \in \mathfrak{R}} R_v$ and since

$v(\xi_0) \geq 0$ for all v in $\mathfrak{R}$ it follows that $\xi_0 \in \bar{S}$, and thus $\xi_0 y_0^{u_0} \in R'$ for some u_0. Similarly $\xi_i y_i^{u_i} \in R'$ for each i. Let $\omega = \xi_0 y_0^q = \xi_i y_i^q$. Then for large μ we have $\omega y_i^{\mu-q} \in R'_\mu$. For large N all monomials in $y_0, \ldots, y_n$ of degree N have this property, and so $\omega R_N \subset R'_{N+q}$ for large N.

Now $(\xi_0) = (\omega/y_0^q) = D - qC_0$. Hence if we can show ω is integral over R, i.e., $\omega \in R'_q$, we can conclude $D \in L'_q$ which will prove the theorem. Let $\zeta = \omega^h$, and let $t = hq$. Then $\zeta R_N \subset R'_{N+t}$, and, if t is large, we have $R'_{N+t} = R_N R'_t$. Therefore $\zeta R_N \subset R'_t R_N$. This shows ζ is integral over $k[R'_t]$ and hence ω is integral over R.

Cor. 12.8: V is arithmetically normal if and only if L_q is complete for all q.

Let R and R' be as before, and let $\mathfrak{L}$ be the conductor of R' in R. $\mathfrak{L}$ is a homogeneous ideal and hence is graded. Therefore we can write $\mathfrak{L} = \mathfrak{L}_0 + \mathfrak{L}_1 + \ldots + \mathfrak{L}_h + \ldots$ where $\mathfrak{L}_0 \neq \emptyset$ if $\mathfrak{L}$ is the unit ideal.

Prop. 12.9: V is normal if and only if $\mathfrak{L}$ contains a power of the irrelevant prime ideal $(y_0, \ldots, y_n)$.

Proof: Assume V is normal. Let ω be any homogeneous element of R'. Let the degree of ω be q. Then ω/y_i^q is integral over $k[V_i]$; and since V is normal, $k[V_i]$ is integrally closed. Therefore $\omega/y_i^q \in k[V_i]$. Hence for some r we have $\omega y_i^r \in k[y]$ which shows that $\omega(y_0, \ldots, y_n)^N \subset k[y]$ for some N. Since R' has a finite basis, there exists an integer h such that $R'(y_0, \ldots, y_n)^h \subset k[y]$. This shows that $(y_0, \ldots, y_n)^h \subset \mathfrak{L}$.

Assume $(y_0, \ldots, y_n)^h \subset \mathfrak{L}$. Let $A_i = k[V_i]$, and let $\zeta \in k(V)$ where ζ is integral over A_i. Then ζy_i^q is integral over $k[y]$ for some q. This shows that $\zeta y_i^{q+h} \in R$ and so ζy_i^{q+h} is a form of degree $q+h$. Therefore $\zeta \in k[V_i]$ which proves $k[V_i]$ is integrally closed and so V is normal.

Prop. 12.10: V is normal if and only if L_q is complete for large q.

Proof: Assume V is normal. Then we have just seen that $(y_o, \ldots, y_n)^h \subset \mathfrak{L}$ for some h. If q is large, $R'_q = R'_{q-h}R_h$ and so $R'_q = R_q$. Therefore $L'_q = L_q$ and L_q is complete.

Assume L_q is complete for large q. Then $R'_q = R_q$ for large q, say $q \geq h$. Therefore $R'(y_o, \ldots, y_n)^h \subset R$ and so $(y_o, \ldots, y_n)^h \subset \mathfrak{L}$. Prop. 12.9 now shows that V is normal.

Let $R = k[y]$, and let $\mathfrak{a}$ be any homogeneous ideal in R. Let the equation of the hyperplane H be $H(Y) = C_oY_o + \ldots + C_nY_n = 0$. Let $V_a = V - V \cap H$. We define

$$\mathfrak{a}_{dh} = \{ f(y_o/H(y), \ldots, y_n/H(y)) \mid f(y) \in \mathfrak{a} \} .$$

Then $\mathfrak{a}_{dh}$ is an ideal in $k[V_a]$, and $\mathfrak{a}_{dh} = (1)$ if and only if $V(\mathfrak{a})$, the variety determined by $\mathfrak{a}$, is contained in H. Under the mapping $\mathfrak{a} \to \mathfrak{a}_{dh}$ prime (primary) ideals go into prime (primary) ideals.

Let V be a variety (affine or projective). Let R be the coordinate ring of V, and let R' be the integral closure of R in its quotient field. Let $\mathfrak{L}(V)$ denote the conductor of R' in R.

Let $P \in V$, and let W be the locus of P over K. Let $\mathfrak{o} = \mathfrak{o}_P(V/k) = \mathfrak{o}_W(V/k)$, and let $\mathfrak{o}'$ be the integral closure of $\mathfrak{o}$. Let $\mathfrak{L}_P(V)$ denote the conductor of $\mathfrak{o}'$ in $\mathfrak{o}$.

The proof of the following proposition is obvious and we omit it.

Prop. 12.11: Let V be a projective variety, and let $\mathfrak{L}(V) = \mathfrak{q}_o \cap \mathfrak{q}_1 \cap \ldots \cap \mathfrak{q}_t$, where

(a) $\mathfrak{q}_o$ is either (1) or is primary for $(y_o, \ldots, y_n)$.

(b) $\mathfrak{q}_i$ is a homogeneous primary ideal ($\mathfrak{p}_i = \sqrt{\mathfrak{q}_i}$).

Let V_a be an affine representative of V and let $P \in V_a$. Then

(1) $\mathcal{L}(V_a) = \bigcap_{i=1}^{t} \mathfrak{q}_{i,dh}$

(2) $\mathcal{L}_P(V) = \bigcap_{i=1}^{t} \mathfrak{o}\,\mathfrak{q}_{i,dh}$

(3) $\mathfrak{q}_{i,dh} = (1)$ if and only if $V(\mathfrak{p}_i)$ is contained in $V - V_a$.

With the conductor $\mathcal{L}$ of V we associate the subvariety $V(\mathcal{L})$ of V. It follows that $V(\mathcal{L}) = \emptyset$ if and only if V is normal.

Cor. 12.12: If $P \in V$, then V is normal at P if and only if $P \notin V(\mathcal{L})$.

Proof: V is normal at P if and only if $\mathfrak{o}' = \mathfrak{o}$. This last condition is equivalent to $\mathcal{L}_P(V) = (1)$ which in turn is equivalent to $P \notin \bigcup_{i=1}^{t} V(\mathfrak{p}_i)$, i.e., to $P \notin V(\mathcal{L})$ ($\mathfrak{p}_i$ is defined as in Prop. 12.11).

Prop. 12.13: If V is a hypersurface (affine or projective) of dimension r, then $\mathcal{L}(V)$ is an unmixed ideal of dimension $r-1$ (both in R and R').

Proof: It will be sufficient to deal with the affine case. Let $f(X_1, \ldots, X_{r+1}) = 0$ be the equation of V. Using a linear transformation with coefficients in the ground field k if necessary, we can assume that $f(X)$ is monic in X_{r+1}. That is, x_{r+1} is integral over $k[x_1, \ldots, x_r]$ where $R = k[x_1, \ldots, x_{r+1}]$.

Let $\mathfrak{p}$ be a prime ideal in R (or R') such that $\dim \mathfrak{p} \leq r-2$. We must show that $\mathcal{L} : \mathfrak{p} = \mathcal{L}$.

Since R and R' are integral over $k[x_1, \ldots, x_r] = A$, $\mathfrak{p} \cap A$ is a prime ideal of the same dimension as $\mathfrak{p}$. Since $\dim \mathfrak{p} \cap A \leq r-2$, $\mathfrak{p} \cap A$ is not principal. Therefore we can find polynomials $g, h \in A \cap \mathfrak{p}$ which are irreducible and relatively prime.

Clearly $\mathcal{L} \subseteq \mathcal{L} : \mathfrak{p}$. Let $w \in \mathcal{L} : \mathfrak{p}$, then $wg(x)$ and $wh(x)$ are in $\mathcal{L}$. Let $\alpha \in R'$. Then $\alpha wg = A(x_1, \ldots, x_{r+1})$ and $\alpha wh = B(x_1, \ldots, x_{r+1})$ are in R where A and B are of smaller degree in

x_{r+1} then the equation of integral dependence of x_{r+1} over $k[x_1,\dots,x_r]$. Since $hA = gB$, this relation must hold for the X_i. Finally, since g and h are not associates, we have $A/g = B/h = c(x_1, \dots, x_{r+1})$ and $\alpha w = c(x)$ is in R. Therefore $w \in \mathfrak{X}$ and so $\mathfrak{X} = \mathfrak{X} : \mathfrak{p}$.

Let V be a variety, let R be its coordinate ring and let R' be the integral closure of R in its quotient field K. If we let K_0 denote $k(V)$, then $K = K_0(y_0)$. Let $\mathfrak{p}'$ be a minimal prime homogeneous ideal in R'. Then $R'_{\mathfrak{p}'}$ is the valuation ring of a prime divisor v' of K. Since t.d. $K/k = r+1$ ($r = \dim_k V$), the residue field of v' has transcendence degree r over k.

Let v be the restriction of v' to K_0. Since $\mathfrak{p}'$ is homogeneous, the v'-residue of y_0 is transcendental over the residue field of v. Therefore the residue field of v has transcendence degree $r-1$ over k, and hence v is a prime divisor of $k(V)$. The center of v on V is the variety of the prime ideal $\mathfrak{p} = \mathfrak{p}' \cap R$ where $\mathfrak{p}$ has the same dimension as $\mathfrak{p}'$. Hence the projective dimension of $\mathfrak{p}$ is $r-1$ which shows that v is a prime divisor of the first kind with respect to V. Conversely, any prime divisor of the first kind with respect to V gives rise to a minimal homogeneous prime ideal in R'. Thus we have a (1-1) correspondence between prime divisors of the first kind with respect to V and minimal homogeneous prime ideals in R'.

Again let $\mathfrak{p}'$ be a minimal homogeneous prime ideal in R', and let Γ be the corresponding prime divisor. Since $\mathfrak{p}'$ is homogeneous, we can write $\mathfrak{p}' = \sum_{m=1}^{\infty} \mathfrak{p}'_m$ where, of course, $\mathfrak{p}'_m \subset R'_m$. Let $LS(\mathfrak{p}'_m)$ be the associated linear system, and let $LS(R'_m) = L'_m$. It is easily seen that $LS(\mathfrak{p}'_m) = L'_m - \Gamma$, i.e., $LS(\mathfrak{p}'_m)$ is the set of Γ-residues of L'_m. Since L'_m is complete, we know by Theorem 5.7 that $LS(\mathfrak{p}'_m)$ is complete.

If $\mathfrak{q}'$ is $\mathfrak{p}'$-primary, then $\mathfrak{q}' = \mathfrak{p}'^{(\nu)}$ for some ν. Hence $\mathfrak{q}'_m = \mathfrak{p}'^{(\nu)}_m$ and $LS(\mathfrak{q}'_m) = L'_m - \nu\Gamma$.

Let $\mathfrak{a}$ be an unmixed homogeneous ideal in R' of projective dimension r-1. Then we can write $\mathfrak{a} = \mathfrak{p}_1'^{(\nu_1)} \cap \mathfrak{p}_2'^{(\nu_2)} \cap \ldots \cap \mathfrak{p}_s'^{(\nu_s)}$ where the $\mathfrak{p}'_i$ are minimal homogeneous prime ideals in R'. Let Γ_i be the prime divisor corresponding to $\mathfrak{p}'_i$. Then $LS(\mathfrak{a}_m) = L'_m - (\nu_1\Gamma_1 + \ldots + \nu_s\Gamma_s)$, and $LS(\mathfrak{a}_m)$ is complete for all m.

In particular, if V is a hypersurface in S_{r+1}, then $\mathfrak{L}$ is a prime (r-1)-dimensional ideal and $LS(\mathfrak{L}_m)$ is complete for all m. If $\{\phi_1(y), \ldots, \phi_h(y)\}$ is a k basis for $\mathfrak{L}_m$, then the $\phi_i(Y)$ are called <u>subadjoint forms</u> of the hypersurface V, and the hypersurfaces H_i determined by $\phi_i(Y) = 0$ are called <u>subadjoint hypersurfaces of</u> V.

§13. The Hilbert characteristic function and the arithmetic genus of a variety.

Let $I = \sum I_m$ be a finitely generated graded module over the polynomial ring $k[Y_o,\dots,Y_n]$. We define the function $\chi(I;m)$ by $\chi(I;m) = \dim_k I_m$ and call $\chi(I;m)$ the characteristic function of I.

An important special case is the one in which $I = k[Y]/\mathfrak{A}$, where $\mathfrak{A}$ is a homogeneous ideal. In this case we shall also write $\chi(\mathfrak{A};m)$ for $\chi(I;m)$.

The following theorem is due to D. Hilbert and J.P. Serre:

> $\chi(I;m)$ is a polynomial in m for large m:
>
> $$\chi(I;m) = \phi_I(m) = a_o\binom{m}{r} + a_1\binom{m}{r-1} + \dots + a_r$$
>
> for large m. Here the constants $a_o, a_1, \dots, a_r$ are necessarily integers, since $\chi(I;m)$ is an integral valued function.

Furthermore, when $I = k[Y]/\mathfrak{A}$ (in which case we write $\phi_{\mathfrak{A}}(m)$ for $\phi_I(m)$), it can be shown that $r = \dim V(\mathfrak{A})$; and if $\mathfrak{A}$ is prime, then a_o is the number of points in which $V(\mathfrak{A})$ is met by a complementary linear variety, i.e., a_o is the order (or degree) of $V(\mathfrak{A})$.

Def. 13.1: If $\mathfrak{A}$ is a homogeneous ideal in $k[Y]$ and if $a_r = \phi_{\mathfrak{A}}(0)$ then $(-1)^r(a_r-1)$ is called the arithmetic genus of the ideal $\mathfrak{A}$ and is denoted by $p_a(\mathfrak{A})$.

Def. 13.2: If V is a variety in S_n, then by the arithmetic genus $p_a(V)$ of V we mean the arithmetic genus of the ideal $\mathfrak{A} = I(V)$.

If $\mathfrak{a} = I(V)$, we may write $\chi(V;m)$, ϕ_V for $\chi(\mathfrak{a};m)$, $\phi_{\mathfrak{a}}$.

Let V be an irreducible r-dimensional variety in S_n, and let $\mathfrak{a} = I(V)$ be the corresponding homogeneous prime ideal. Let $R = k[Y]/\mathfrak{a}$ be the homogeneous coordinate ring of V. Then $\chi(V;m) = \dim_k R_m = 1 + \dim L_m$, and for large m

$$\chi(V;m) = a_0\binom{m}{r} + a_1\binom{m}{r-1} + \dots + a_{r-1}\binom{m}{1} + a_r.$$

Let R' be the integral closure of R. Then $\chi(R';m) = \dim_k R'_m = 1 + \dim L'_m$, and for large m

$$\chi(R';m) = \phi_{R'}(m) = a'_0\binom{m}{r'} + a'_1\binom{m}{r'-1} + \dots + a'_{r'}.$$

We shall show that $r' = r$ and $a_0 = a_0'$. Let $\psi(y)$ be in the conductor $\mathfrak{c}$ of R' in R where $\psi(y)$ is homogeneous of degree h. Then $\psi(y)R'_m \subset R_{m+h}$, and, of course, $R_m \subset R_m'$. Hence $\phi_R(m) \leq \phi_{R'}(m) \leq \phi_R(m+h)$, for h a fixed integer and for all large m. Therefore ϕ_R and $\phi_{R'}$ have the same degree and the same leading coefficients, i.e., $r = r'$ and $a_0/r! = a_0'/r!$. Hence $a_0 = a_0'$.

Let q be large enough so that $V' = V'^q$ is arithmetically normal. Then $\chi(V';m) = 1 + \dim L'_{mq} = \chi(R';mq)$. Hence $\phi_{V'}(m) = \phi_{R'}(mq)$, i.e., $\phi_{V'}(m) = a_0'q^r\binom{m}{r} + \dots + a'_r$. Thus $\phi_{V'}(0) = \phi_{R'}(0)$. Therefore, for large q, all derived normal models V'^q of V have the same arithmetic genus, namely $(-1)^r(a'_r - 1)$.

Let V be a normal variety, and let D be an effective divisor cycle on V, i.e., $D = m_1\Gamma_1 + \dots + m_h\Gamma_h$, $m_i > 0$ for all i.

Let $\mathfrak{p}_i = I(\Gamma_i)$ in $R = k[V]$. Let $\mathfrak{a} = \mathfrak{p}_1^{(m_1)} \cap \mathfrak{p}_2^{(m_2)} \cap \ldots \cap \mathfrak{p}_h^{(m_h)}$. We wish to study $\chi(\mathfrak{a};m) = \chi(R/\mathfrak{a};m)$, and we write $\chi(D;m)$ for $\chi(\mathfrak{a};m)$ and ϕ_D for $\phi_{\mathfrak{a}}$.

Def. 13.3: The arithmetic genus $p_a(D)$ is $p_a(\mathfrak{a})$, i.e., $(-1)^{r-1}[\phi_D(0) - 1]$.

Prop. 13.4: $\chi(D;m) = \dim L_m - \dim(L_m - D)$ for all m, and consequently $\phi_D(m) = \dim L_m - \dim(L_m - D)$ for large m.

Proof: Let $I = R/\mathfrak{a}$. Then $I_m = R_m/\mathfrak{a}_m$ and therefore $\chi(D;m) = \dim_k R_m - \dim_k \mathfrak{a}_m$. We know that $\dim_k R_m = 1 + \dim L_m$. Since $LS(\mathfrak{a}_m) = L_m - D$, we have $\dim_k \mathfrak{a}_m = 1 + \dim(L_m - D)$ which proves the first part and hence the proposition.

Cor. 13.5: If $D \equiv D'$ where D and D' are both effective, then $\phi_D = \phi_{D'}$, and, in particular, $p_a(D) = p_a(D')$.

Proof: If m is large, $L_m - D$ is a complete system. Hence

$$L_m - D = |L_m - D| = L_m - D'.$$

Prop. 13.6: If X and Y are effective divisorial cycles on a non-singular surface, then $\phi_X + \phi_Y = \phi_{X+Y} + (X.Y)$.

Proof: In view of Cor. 13.5, we may assume X and Y have no common component.

Let W_1 and W_2 be subspaces of a finite-dimensional vector space W. Then it is well known that $\dim W_1 + \dim W_2 = \dim(W_1 + W_2) + \dim(W_1 \cap W_2)$. Now let A and B be homogeneous ideals in $R = k[y_0, \ldots, y_n]$. Let A_m, B_m denote the spaces of forms of degree m in A and B respectively. Then $\dim A_m + \dim B_m = \dim(A+B)_m + \dim(A \cap B)_m$. Since $\chi(A;m) + \dim A_m = \dim R_m$, etc., we find

$$\phi_{A+B} + \phi_{A\cap B} = \phi_A + \phi_B.$$

Let $A = I(X)$ and $B = I(Y)$. Then $I(X+Y) = A \cap B$. By definition $\phi_A = \phi_X$, $\phi_B = \phi_Y$ and $\phi_{A\cap B} = \phi_{X+Y}$. Hence it will be sufficient to prove that $\phi_{A+B} = (X.Y)$.

Let $P_1, \ldots, P_t$ be the common points of $[X]$ and $[Y]$, and let $\mathfrak{P}_i = I(P_i)$. Since $V(A+B) = [X]\cap[Y]$, we have $A+B = \mathfrak{q}_0 \cap \mathfrak{q}_1 \cap \ldots \cap \mathfrak{q}_t$ where $\mathfrak{q}_i$ is $\mathfrak{P}_i$-primary for $1 \leq i \leq t$ and $\mathfrak{q}_0$ is primar for the irrelevant prime ideal $(y_0, \ldots, y_n)$. We have $\mathfrak{q}_0 \supset (y_0, \ldots, y_n)^h$ for some integer h, i.e., all forms in the y_i of degree $\geq h$ are in $\mathfrak{q}_0$. Hence $\phi_{A+B} = \phi_{\mathfrak{q}_1 \cap \ldots \mathfrak{q}_t}$. Let $Q_i = \bigcap_{\substack{j=1 \\ j\neq i}}^{t} \mathfrak{q}_i$. Then $V(\mathfrak{q}_i + Q_i) = \emptyset$ and so $\phi_{\mathfrak{q}_i + Q_i} = 0$. Therefore, again using the dimension argument, we have $\phi_{A+B} = \phi_{\mathfrak{q}_1} + \ldots + \phi_{\mathfrak{q}_t}$ where the $\phi_{\mathfrak{q}_i}$ are all constants since $\dim V(\mathfrak{q}_i) = 0$. We shall show $\phi_{\mathfrak{q}_j} = i(X,Y;P_j)$.

Let P be any of the P_j, and let $\xi = 0$, $\eta = 0$ be local equations of X and Y respectively at P. Then $i(X,Y;P) = \dim \mathcal{O}_P/\mathcal{O}_P(\xi,\eta) = \nu$. Let $\{\omega_1, \ldots, \omega_\nu\}$ be a k-basis of $\mathcal{O}_P$ mod $\mathcal{O}_P(\xi,\eta)$. Since the elements of $\mathcal{O}_P(\xi,\eta)$ are quotients $F(y)/G(y)$ of forms of the same degree where $G(P) \neq 0$ and $F(y) \in \mathfrak{q}$, we can choose $\omega_i = f_i(y)/y_0^{\mu}$ where the $f_i(y)$ are forms of degree μ. To show $\phi_{\mathfrak{q}} = i(X,Y;P)$, it is sufficient to show that $\phi_{\mathfrak{q}}(m) = \nu$ for large m. This is an obvious consequence of the following two simple assertions:

(1) $f_1(y),\ldots,f_\nu(y)$ are linearly independent over k(mod $\mathfrak{P}$).

(2) If $m \geq \mu$, then the forms $y_0^{m-\mu} f_j(y)$, $j = 1,\ldots,\nu$ form a basis of R_m (mod $\mathfrak{P}$).

<u>Cor. 13.7</u>: If X and Y are effective divisorial cycles on a non-singular surface, then $p_a(X+Y) = p_a(X) + p_a(Y) + (X.Y) - 1$.

In §11 we gave a different definition of the arithmetic genus. Now we want to show that the two definitions are equivalent. If V is an irreducible curve, we shall denote by $\pi(V)$ the effective genus of V. If D is a divisorial cycle on our surface F, we denote by p(D) the arithmetic genus of D, as defined in §11. We first prove the following :

<u>Prop. 13.8</u>: Let V be an irreducible curve in S_n, and let $\phi_V(m) = a_0 m - p_a(V) + 1$. Then $p_a(V) \geq \pi(V)$, and $p_a(V) = \pi(V)$ if and only if V is non-singular

Proof: We have $\phi_V(m) = 1 + \dim L_m$, hence $\dim L_m = a_0 m - p_a(V)$. If V' is a normalization of V, then $\dim L_m' = a_0 m - p_a(V')$. Let ν be the degree of the divisors in L_1 (hence ν is the order of V). Then the divisors in L_m' have degree $m\nu$. The Riemann-Roch theorem shows that $\dim L_m' = m\nu - \pi(V)$. Therefore $\nu = a_0$ and $p_a(V') = \pi(V)$. Since $L_m \subset L_m'$, we have $p_a(V) \geq p_a(V') = \pi(V)$.

The last assertion is a consequence of the obvious equivalence of the following statements:

(a) $p_a(V) = \pi(V)$.

(b) $L_m = L_m'$ for large m.

(c) V is normal.

(d) V is non-singular (since V is a curve).

From this proposition and from §11 it follows that $p_a(V) = p(V)$ for any irreducible non-singular curve on F. From Prop. 11.3(a) and above Cor. 13.7 it follows that $p_a(D) = p(D)$ for any curve D on F (i.e., any effective divisorial cycle D all components of which enter in D with coefficient 1) such that each irreducible component of D is non-singular. We now can sketch the proof of the general

Prop. 13.9: If D is an effective divisorial cycle on a non-singular surface F, then $p_a(D) = p(D)$.

First we assert without proof that the system L_1 of hyperplane sections of F contains curves all components of which are non-singular curves. This implies that if $C_1 \in L_1$, then $p_a(C_1) = p(C_1)$. Thus if $C_m \in L_m$, then $p_a(C_m) = p(C_m)$.

Let q be large enough so that there exist effective cycles E satisfying $E+D \in L_q$. Then $|L_m+D| = |L_{m+q} - E|$. We shall also assume without proof a special case of the theorem of Bertini: If E is an effective cycle and m is large, then the system $|L_m-E|$ contains non-singular curves. Let Z be a cycle in $|L_{m+q}-E|$ all of whose components are non-singular curves with multiplicity one. Then $Z \in |L_m+D|$. Hence there exists $C_m \in L_m$ such that $C_m + D \equiv Z$. Since the proposition holds for C_m and Z, it must hold for D.

<u>Prop. 13.10</u>: If D is any effective cycle on F, then $\phi_D(m) = (D.C_m) - p(D) + 1$ where C_m is any cycle in L_m.

Proof: Assume $D = X+Y$ where $X \geq 0$ and $Y \geq 0$. Then $\phi_D(m) = \phi_X(m) + \phi_Y(m) - (X.Y)$. If the proposition is true for X and Y, then

$$\begin{aligned}\phi_D(m) &= (X.C_m) - p(X) + 1 + (Y.C_m) - p(Y) + 1-(X.Y)\\ &= (D.C_m) - \{ p(X) + p(Y) + (X.Y) - 1 \} + 1\\ &= (D.C_m) - p(X+Y) + 1\\ &= (D.C_m) - p(D) + 1.\end{aligned}$$

Thus we need only prove the proposition for an irreducible curve D

For an irreducible curve D, $\phi_D(m) = \nu m - p(D) + 1$ where ν is the order of D. Let L_m be the system cut out on F by hypersurfaces of order m, and let L_m^* be the system cut out on D by hypersurfaces of order m. Let $(y_o,\ldots,y_m)$ be a general point of F, and let $(y_o^*,\ldots,y_m^*)$ be a general point of D. Finally, let $R = k[y]$, and let $R^* = k[y^*]$. We have $L_m = LS(R_m)$ and $L_m^* = LS(R_m^*)$. Hence $L_m^* = Tr_D L_m$. Therefore, if C_m^* is a divisor in L_m^*, then $C_m^* = D.C_m$ for some C_m in L_m. Since ν is the degree of the divisors in L_1^*, νm is the degree of the divisors in L_m^*. Hence $\nu m = (D.C_m)$.

<u>Prop. 13.11</u>: If D is any effective cycle, then

$$p(-D) = (D^2) - p(D) + 2 = p(D-K)$$

where K is a canonical cycle.

Proof: The first equality follows from

$$1 = p(0) = p(D + (-D)) = p(D) + p(-D) - (D^2) - 1.$$

By definition (see the beginning of §11) we have

$$-(D.K) = (D^2) - 2p(D) + 2.$$

The second equality now follows from $p(D-K) = p(D) + p(-K) - (D.K) - 1$ and from $p(-K) = 1$.

Cor. 13.12: $p(-D') = p(D)$ where $D' \equiv D + K$.

Theorem 13.13: For any divisorial cycle D of F there exists an integer N(D) such that for all $m \geq N(D)$ we have

$$\dim|D + C_m| = p_a(F) + p(-D-C_m) - 1.$$

Proof: Assume first $D = 0$. In this case we must show that for large m $\dim L_m = p_a(F) + p(-C_m) - 1$. We know that for large m $\dim L_m = \phi_F(m) - 1$. We assert that $\phi_F(m) = p_a(F) + p(-C_m)$ for all m. If $m = 0$, we know $\phi_F(0) = 1 + p_a(F)$. Since, for large m, $\dim L_m - \dim(L_m-D) = \phi_D(m)$, and since $L_{m-1} = L_m-C_1$, it follows that $\phi_F(m) - \phi_F(m-1) = \phi_{C_1}(m)$. Our assertion now follows by induction on m. Thus the theorem is true for $D = 0$.

There exists an integer m such that L_m-D contains an effective cycle E. Then $D + C_m \equiv C_{m+q} - E$. Let $Z = D + C_m$, then $E \equiv C_{m+q} - Z$. We have already observed that $\dim|C_{m+q}-E| = \dim L_{m+q} - \phi_E(m+q)$. By Prop. 13.10 $\phi_E(m+q) = (E.C_{m+q}) - p(E) + 1$. Hense

$$\phi_E(m+q) = (C_{m+q}^2) - p(C_{m+q}) - p(-Z) + 2$$

by Cor. 13.7. Therefore

$$\dim|D+C_m| = \dim|C_{m+q}-E| = \dim L_{m+q}-(C_{m+q}^2) + p(C_{m+q}) + p(-Z) - 2.$$

Since the theorem if true for $D = 0$, we see that

$$\dim |C_{m+q}| = \dim L_{m+q} = p_a(F) + p(-C_{m+q}) - 1.$$

Thus

$$\dim |D+C_m| = p_a(F) + p(-C_{m+q}) - (C_{m+q}^2) + p(C_{m+q}) + p(-Z) - 3.$$

Now applying Prop. 13.11 we see that

$$\dim |D + C_m| = p_a(F) + p(-Z) - 1$$

which proves the theorem.

Remark. The expression $p_a(F) + p(-Z) - 1$ is a numerical character of cycles Z. Let us denote it by $\rho(Z)$. If Z is of the form $D + C_m$, where D is a given cycle and m is sufficiently large ($m \geqq N(D)$, where $N(D)$ depends on D), then $\rho(Z)$ gives the dimension of $|Z|$. In all cases, $\rho(Z)$ is sometimes referred to as the virtual dimension of $|Z|$. A simple calculation leads to the following alternative expressions of $\rho(Z)$:

$$\begin{aligned}\rho(Z) &= p_a(F) + p(-Z) - 1\\ &= (Z^2) - p(Z) + p_a(F) + 1\\ &= \tfrac{1}{2}[(Z^2) - (K.Z)] + p_a(F)\\ &= \bar{p}(Z) - (K.Z) + p_a(F) - 1.\end{aligned}$$

§14. The Riemann-Roch Theorem.

Let F be a non-singular surface and D a divisorial cycle on F. Let K be a canonical divisor on F. We define i(D), the <u>index of specialty of</u> D, to be

$$i(D) = 1 + \dim |K-D|.$$

In this section we shall denote the geometric genus of F by p_g. If $p_g = 0$, then $|K|$ does not exist, nor does $|K-D|$ if D is effective. This shows that if $p_g = 0$ and D is effective, then $i(D) = 0$.

The Riemann-Roch theorem says that

$$\dim |D| \geq p_a(F) + p_a(-D) - 1 - i(D).$$

<u>Def. 14.1</u>: We define the <u>deficiency</u> of a cycle Z, denoted by $\delta(Z)$, by $\delta(Z) = p_g + p(Z) - 1 - \dim |Z+K|$, where p(Z) is the arithmetic genus of Z.

The following is a restatement of some earlier results.

<u>Cor. 14.2</u>: If E is an irreducible non-singular curve, then $\delta(E)$ is the deficiency of the system $Tr_E|E+K|$, i.e.,

$$\delta(E) = p(E) - 1 - \dim Tr_E|E+K|.$$

<u>Prop. 14.3</u>: If $C \equiv D+E$, where E is an irreducible non-singular curve, then $\delta(D) \leq \delta(C)$ provided $(D.E) > 0$.

Proof: Let $C' = C+K$, $D' = D+K$ and $E' = E+K$. Then $C' \equiv D'+E = D+E'$. Hence we have

$$\dim |C'| = \dim |D'| + \dim \mathrm{Tr}_E|C'| + 1 \qquad (1)$$

by Cor. 7.12. Fix Z such that $Z \equiv C'$ and E is not a component of Z. We can consider Z.E as a cycle on E, hence $\mathrm{Tr}_E|C'|$ is a subsystem of the complete system $|Z.E|$ on E. The degree of the divisors in $|Z.E|$ is $(Z.E) = (C'.E) = (E'.E) + (D.E) = 2p(E) - 2 + j$ where $j = (D.E)$. Since $j > 0$ by assumption, we have $(Z.E) > 2p(E) - 2$. Thus we can apply the Riemann-Roch theorem for curves to E obtaining the result $\dim |Z.E| = p(E) - 2 + j$. Hence $\dim \mathrm{Tr}_E|C'| \leq p(E) - 2 + j$. Using (1) we have

$$\dim |C'| \leq \dim |D'| + p(E) - 1 + j. \qquad (2)$$

Def. 14.1 yields $\dim |C'| = p_g + p(C) - 1 - \delta(C)$ and

$$\dim |D'| = p_g + p(D) - 1 - \delta(D).$$

Hence $\delta(D) - \delta(C) \leq p(D) + p(E) + (E.D) - 1 - p(C)$. Since $p(C) = p(D) + p(E) + (E.D)-1$, we have $\delta(D) - \delta(C) \leq 0$, as asserted.

We shall now assume the following special case of Bertini's Theorem:

For any $m \geq 1$, there exist non-singular irreducible curves in L_m.

<u>Cor. 14.4</u>: If C_m denotes a cycle in L_m, then $\delta(C_i) \leq \delta(C_{i+1})$ for $i = 1,2,3,\ldots$

Proof: If $C_m \in L_m$, we can find a $C_{m-1} \in L_{m-1}$ and an irreducible non-singular C_1 in L_1 such that $C_m \equiv C_{m-1} + C_1$ and $(C_m.C_1) = m(C_1^2)$. Since $(C_1^2) > 0$, the result now follows from Prop. 14.3.

<u>Prop. 14.5</u>: If m is large, then $\delta(C_m) = p_g - p_a$ where $p_a = p_a(F)$.

Proof: Let $C'_m = C_m + K$. By definition of $\delta(C_m)$, $\dim |C'_m| = p_g + p(C_m) - 1 - \delta(C_m)$. Since m is large, Theorem 13.13 allows us to conclude that $\dim |C'_m| = p_a + p(-C'_m) - 1$. Since $p(-C'_m) = p(C_m)$ [see Cor. 13.12], $\dim |C'_m| = p_a + p(C_m) - 1$. The result now follows.

<u>Theorem 14.6</u>: If D is any cycle such that $(D.C_1) > 0$, then $\delta(D) \leq p_g - p_a$.

Proof: Consider $|L_m - D|$ where m is large. We shall assume known the fact that for sufficiently large m, $|L_m - D|$ contains irreducible non-singular curves E. Let $C_m \equiv D+E$. Then $(D.C_m) = m(D.C_1) \geq m$. On the other hand, $(D.C_m) = (D^2) + (D.E) \geq m$. Hence if m is large, we must have $(D.E) > 0$. Therefore $\delta(D) \leq \delta(C_m)$ by Prop. 14.3. The theorem now follows from Prop. 14.5.

<u>Remark</u>: By definition, $\delta(D) = p_g + p(D) - 1 - \dim |D'|$, where $D' = D+K$. Hence Theorem 14.6 says that $p(D) - 1 - \dim |D'| \leq -p_a$ or $p(D) + p_a - 1 \leq \dim |D'|$. Since $p(D) = p(-D')$, this means that $p(-D') + p_a - 1 \leq \dim |D'|$. Now $i(D') = 1 + \dim |K-D'| = 1 + \dim |-D|$, and thus $i(D') = 0$ since the intersection number of $-D$ with C_1 is negative and since therefore $|-D|$ must be empty. Thus $\dim |D'| \geq p(-D') + p_a - 1 - i(D')$. Hence Theorem 14.6 is the Riemann-Roch theorem for the special systems of the form $|D+K|$ where $(D.C_1) > 0$.

The rest of these lectures will be devoted to the proof of the theorem of Riemann-Roch, or rather to the proof of a fundamental lemma (Enriques-Severi-Zariski) on which the proof of the Riemann-Roch theorem is based.

Theorem of Riemann-Roch. If F is a non-singular surface and D is any divisorial cycle on F, then

$$\dim |D| = p_a + p(-D) - i(D) + \delta - 1$$

where $\delta \geq 0$.

Proof: By Cor. 7.12 we have $\dim |D| = \dim |D+C_m| - \dim Tr_{C_m} |D+C_m| - 1$, where C_m is an irreducible non-singular curve in L_m which is not a component of D. Let $d = \dim Tr_{C_m} |D+C_m|$. Then Theorem 13.13 asserts that $\dim |D| = p_a + p(-D - C_m) - d - 2$ for large m.

Let Z be a fixed cycle such that $Z \equiv D + C_m$ and C_m is not a component of Z. Then we can again view $Z.C_m$ as a cycle on C_m. As such it is an element of $Tr_{C_m} |D+C_m|$. Hence $Tr_{C_m} |D+C_m| \subseteq |Z.C_m|$ This allows us to write $d = \dim |Z.C_m| - \delta$ where $\delta \geq 0$. Since C_m is irreducible and non-singular, the Riemann-Roch theorem for curves yields $\dim |Z.C_m| = (Z.C_m) - p(C_m) + \dim |K(C_m) - Z.C_m| + 1$, where $p(C_m) = \pi(C_m)$, the genus of C_m, and where $K(C_m)$ is a canonical divisor of C_m.

Let K be a canonical divisor on F such that C_m is not a component of K. Let $C_m^{(1)} \in L_m$ where $C_m^{(1)} \neq C_m$. Then by Prop. 11.1 we can take $K.C_m + C_m^{(1)}.C_m$ as our canonical cycle $K(C_m)$. Furthermore, since C_m is not a component of either D or

$C_m^{(1)}$, we can take $D+C_m^{(1)}$ as our cycle Z. Then $K(C_m)-Z.C_m = (K-D).C_m$, and $\dim |Z.C_m| = (D.C_m) + (C_m^2) - p(C_m) + \dim |(K-D).C_m| + 1$. Hence $\dim |D| = p_a + p(-D-C_m) - (D.C_m) - (C_m^2) + p(C_m) - \dim |(K-D).C_m| + \delta - 3$.

Since $p(-D) = p(-D-C_m+C_m) = p(-D-C_m) + p(C_m) - (D.C_m) - (C_m^2) - 1$, we have

$$\dim |D| = p_a + p(-D) - \dim |(K-D).C_m| + \delta - 2. \qquad (*)$$

To complete our proof we need the following result which we shall refer to as the Fundamental Lemma:

<u>Fundamental Lemma</u>: For any divisorial cycle D there exists an integer N(D) such that the $Tr_{C_m}|D|$ is complete if $m \geq N(D)$. More precisely: if D_1 is any cycle linearly equivalent to D such that C_m is not a component of D_1, then $Tr_{C_m}|D|$ coincides with the complete system $|D_1.C_m|$ on C_m.

For the moment let us assume the Fundamental Lemma and complete the proof of the Riemann-Roch theorem. If m is large, then, by the Lemma, $Tr_{C_m}|K-D|$ is complete. Therefore we have $\dim |(K-D).C_m| = \dim Tr_{C_m}|K-D| = \dim |K-D| - \dim |K-D-C_m| - 1$. If m is large, then $|C_m+D-K|$ contains strictly positive cycles. Hence $\dim |K-D-C_m| = -1$ and so $\dim Tr_{C_m}|K-D| = \dim |K-D|$. Substituting this in (*) we have

$$\dim |D| = p_a + p(-D) - \dim |K-D| + \delta - 2.$$

By definition $i(D) = \dim |K-D| + 1$. Hence

$$\dim |D| = p_a + p(-D) - i(D) + \delta - 1, \qquad \text{Q.E.D.}$$

We now turn to the proof of the F.L. (Fundamental Lemma).

Prop. 14.7: If the F.L. is true for a given cycle D, then it is true for any cycle $D_1 \leq D$.

To prove this we assume for a moment the following lemma:

Lemma 14.8: Let E be an effective cycle on F, and let $|D|$ be a given complete linear system. There exists an integer $M = M(D,E)$ such that if $D' \in |D|$ and if $D'.C_m \geq E.C_m$ for some $m \geq M$, then $D' \geq E$.

Proof of 14.7: Let $\overline{Z} \in |D_1.C_m|$ (we assume that C_m is not a component of D). Since $D_1 \leq D$, we can write $D = D_1+E$ where E is effective. Hence $D.C_m \equiv \overline{Z} + E.C_m$. We take $m \geq \max.\{N(D),M(D,E)\}$. Since $m \geq N(D)$, then $\overline{Z} + E.C_m \in Tr_{C_m}|D|$ because the F.L. is true for D. Hence there exists $D^{(1)} \in |D|$ such that

(a) $D^{(1)}.C_m$ is defined, and

(b) $D^{(1)}.C_m = \overline{Z} + E.C_m$.

Since $\overline{Z}$ is effective, we have $D^{(1)}.C_m \geq E.C_m$. Applying Lemma 14.8, we have $D^{(1)} = E+D'$ where $D' \geq 0$ (since $m \geq M(D,E)$). Since $D = D_1+E$ and $D^{(1)} \in |D|$, we see that $D' \equiv D_1$. Hence $D' \in |D_1|$ and $D'.C_m = \overline{Z}$. Thus $\overline{Z} \in Tr_{C_m}|D_1|$ and so $|D_1.C_m| = Tr_{C_m}|D_1|$. Hence the F.L. holds for D_1.

Proof of 14.8: Let Γ be a prime component of E, of multiplicity h. Let s be the order of the curve Γ, i.e., $s = (\Gamma.C_1)$ where C_1 is a plane section. Let m be chosen so that $ms \geq (D'.\Gamma) + h|(\Gamma'^2)|$. We shall show that this implies $D' \geq h\Gamma$. If we assume this for the moment, then by applying this result to the other components of E we obtain a lower bound for m and the proposition is proved. Hence we must show that $D' \geq h\Gamma$ if $ms \geq (D'.\Gamma) + h|(\Gamma^2)|$.

Let ν be the multiplicity of Γ in D'. Since D' is effecitve, $\nu \geq 0$. We must show $\nu \geq h$. Assume $\nu < h$. Let $X = D' - \nu\Gamma$. Then X is an effective cycle, and Γ is not a component of X. By our choice of m we see that

$$ms > (X.\Gamma) . \tag{1}$$

We can write $D'.C_m = E.C_m + \overline{Z}$ where $\overline{Z}$ is an effective cycle on C_m. Hence $X.C_m + \nu\Gamma.C_m = h\Gamma.C_m + (E-h\Gamma).C_m + \overline{Z}$. Therefore

$$X.C_m \geq (h-\nu)\Gamma.C_m . \tag{2}$$

Let $P \in \Gamma \cap C_m$. Since C_m is a non-singular curve, we can choose x and y as uniformizing parameters at P such that y = 0 is a local equation of C_m at P. Let $\xi = 0$ and $\gamma = 0$ be local equations of X and Γ respectively at P. Let $\sigma = i(\Gamma, C_m ; P)$ and $\sigma' = i(X, C_m ; P)$. By (2) we have $\sigma' \geq \sigma$. We may write $\gamma = Ay + Bx^{\sigma}$ and $\xi = Cy + Dx^{\sigma'}$ where A, B, C, and D are in $\mathcal{O} = \mathcal{O}_P(F/k)$ and $B, D \notin \mathfrak{m}$. By definition of intersection multiplicity, we have:

$$i(X, \Gamma ; P) = \dim_k \mathcal{O}/\mathcal{O}(\xi, \gamma) \geq \dim_k \mathcal{O}/\mathcal{O}(y, x^{\sigma}) = \sigma = i(\Gamma, C_m ; P),$$

where the inequality follows from $\mathcal{O}(\xi, \gamma) \subset \mathcal{O}(y, x^{\sigma})$. Therefore $(X.\Gamma) \geq (\Gamma.C_m) = ms$. Since this contradicts (1), we must have $\nu \geq h$, Q.E.D.

In view of Prop. 14.7 it is sufficient to prove the F.L. for any collection L of divisorial cycles D_i having the property that given any divisorial cycle Z there exists a cycle D_i in the collection such that D_i-Z is linearly equivalent to an effective cycle. We may say then that L is cofinal with the totality of all divisoria cycles (up to linear equivalence). Now, if D is any fixed divisoria cycle then the cycles $D_i = D+C_i$ form such a collection L. In our paper, "Complete linear systems on normal varieties and a generaliza tion of a lemma of Enriques-Severi" (Ann. of Math., 1952) we have taken for L the collection of cycles C_i, $i = 1,2,\ldots$ $(D = 0)$. In these lectures we shall take as fixed cycle D a canonical cycle K. In other words, we shall prove that for each $i = 1,2,\ldots$ there exists an integer $N(i)$ such that $Tr_{C_m}|K+C_i|$ is complete if $m \geqq N(i)$. This is precisely the original formulation of the fundamental lemma, due to Severi. (See our monograph, "Algebraic Surfaces", p. 67). Our present proof of the F.L. for algebraic surfaces (which can be easily extended to varieties) makes use of regular differentials of degree 2 and is based on the precise relationships (which will be established in the next section) between the geometric concepts of an adjoint and subadjoint surface of a surface in S_3 and such arithmetic concepts as conductor, different, and complementary module. As in the classical proofs of the Italian geometers, we shall also have to project our non-singular surface F into a surface F_0 in S_3, birationally equivalent to F. However, we shall have no need of the proposition that the projection can be made in such a fashion that F_0 has only "ordinary" singularities. All we shall require is that the birational correspondence between F and F_o have no fundamental points on either surface; in other words: that F be a normalization of F_o. The existence of a projection F_o satisfying this condition is an immediate consequence of the "normalization theorem" of Emmy Noether

§15. Subadjoint polynomials.

Let S be an integrally closed noetherian domain, and let L be its quotient field. Let K be a finite, separable, algebraic extension of L. Finally, let R be a ring such that $S \subset R \subset K$ and such that

(1) R is integral over S, and

(2) K is the quotient field of R.

We define the complementary module $\mathcal{C}_{R/S} = \{ z \in K \mid Tr_{K/L}\, zu \in S$ for all $u \in R \}$. Clearly $\mathcal{C}_{R/S}$ is a module over both S and R. Furthermore it is a finitely generated module; and since S is integrally closed, we have $R \subset \mathcal{C}_{R/S}$. By the different $\mathfrak{D}_{R/S}$ we mean the set $R : \mathcal{C}_{R/S} = \{ \xi \in K \mid \xi \mathcal{C} \subset R \}$. $\mathfrak{D}_{R/S}$ is an ideal in R (since $1 \in \mathcal{C}$).

Assume R = S[y] where y is a primitive element of K/L and y is integral over S. Let f(Y) be the (monic) minimal polynomial of y over L (whence $f(Y) \in S[Y]$). It is known that in this case $\mathcal{C}_{R/S} = (1/f'(y)S[y]$ and hence $\mathfrak{D}_{R/S} = f'(y)S[y]$.

Assume S is a Dedekind domain, and let K be as above. Let R' be the integral closure of S in K. Since K/L is separable, R' is also a Dedekind domain. Let $\mathfrak{C} = \mathfrak{C}_{R/S}$, $\mathfrak{C}' = \mathfrak{C}_{R'/S}$, $\vartheta = \vartheta_{R/S}$, $\vartheta' = \vartheta_{R'/S}$, and let $\mathfrak{L}$ be the conductor of R' in R. Then it can be shown that $R'\vartheta \subset \mathfrak{L}\vartheta'$ and that if $\mathfrak{C}$ is invertible, i.e., if $\mathfrak{C}\vartheta = R$, then $R'\vartheta = \mathfrak{L}\vartheta'$.

<u>Prop. 15.1</u>: Let K/k = k(V) where dim V = r, K/k is separably generated and k is an arbitrary ground field which is maximally algebraic in K. Let $\{x_1,\dots,x_r\}$ and $\{y_1,\dots,y_r\}$ be two separating transcendence bases of K/k. Let $\mathfrak{p}$ be a prime divisor of K/k where $\mathfrak{p}$ is of the first kind with respect to k[x] and k[y]. Let R_x be the integral closure of k[x] in K, and let R_y be the integral closure of k[y] in K. Finally, let $\vartheta_x = \vartheta_{R_x/k[x]}$ and let $\vartheta_y = \vartheta_{R_y/k[x]}$.

Then $v_{\mathfrak{p}}\left(\dfrac{\partial(y_1,\dots,y_r)}{\partial(x_1,\dots,x_r)}\right) = v_{\mathfrak{p}}(\vartheta_y) - v_{\mathfrak{p}}(\vartheta_x)$.

Proof: We proceed by induction on r, since for r = 1 the result is known (see, for instance, "Algebraic Functions of One Variable", by C. Chevalley).

We may assume $\mathfrak{p}$ is trivial on $k^* = k(x_2,\dots,x_r)$. We may also assume that <u>each</u> y_i $(i = 1,2,\dots,r)$ has a transcendental $\mathfrak{p}$-residue, because if, say, $\mathfrak{p}$ is trivial on $k(x_2,x_3,\dots,x_r)$ and if y_1 has an algebraic $\mathfrak{p}$-residue, we can replace it by $y_1 + y_2$ (which has a transcendental $\mathfrak{p}$-residue) without changing the Jacobian.

We know that t.d. $K/k^* = 1$ and x_1 is a separating transcendental. Clearly $k^*(y_1,\ldots,y_r)/k^*$ is a separably generated extension of transcendence degree one, and $K/k^*(y_1,\ldots,y_r)$ is a separable algebraic extension. Let y_1 be a separating transcendental of $k^*(y_1,\ldots,y_r)/k^*$. Then y_1 is also a separating transcendental of K/k^*. Hence $\{y_1,x_2,\ldots,x_r\}$ is a separating transcendence basis of K/k.

We adopt the following notations:

(1) $k_1 = k(y_1)$.

(2) $R_{x,y}$ is the integral closure of $k[y_1,x_2,\ldots,x_r]$ in K.

(3) $\vartheta_{x,y} = \vartheta_{R_{x,y}/k[y_1,x_2,\ldots,x_r]}$.

Furthermore we know that

(4) x_1 and y_1 are both separating transcendentals of K/k^*, and

(5) $\{x_2,\ldots,x_r\}$ and $\{y_2,\ldots,y_r\}$ are separating transcendence bases of K/k_1.

Derivations in K/k^* will be denoted by D^*, and those in K/k_1 by D_1. We have

$$D^*_{x_1} y_1 = \frac{\partial(y_1,x_2,\ldots,x_r)}{\partial(x_1,x_2,\ldots,x_r)} \quad \text{and} \quad \frac{\partial_1(y_2,\ldots,y_r)}{\partial_1(x_2,\ldots,x_r)} = \frac{\partial(y_1,y_2,\ldots,y_r)}{\partial(y_1,x_2,\ldots,x_r)}$$

Hence

$$\frac{\partial(y_1,y_2,\ldots,y_r)}{\partial(x_1,x_2,\ldots,x_r)} = \frac{\partial_1(y_2,\ldots,y_r)}{\partial(x_2,\ldots,x_r)} \cdot D^*_{x_1} y_1 . \tag{1}$$

Let M denote the multiplicative set $k[x_2,x_3,\ldots,x_r] - \{0\}$, let $R_{x;M}$ and $R_{x,y;M}$ be the quotient rings, with respect to M, of R_x and $R_{x,y}$ respectively, and let $\vartheta_{x;M} = R_{x;M} \cdot \vartheta_x$,

$\mathcal{D}_{x,y;M} = R_{x,y;M} \cdot \mathcal{D}_{x,y}$. Then it is immediately seen that $R_{x;M}$ is the integral closure of $k^*[x_1]$ in K, $R_{x,y;M}$ is the integral closure of $k^*[y_1]$ in K and that

$$\mathcal{D}_{x;M} = \mathcal{D}_{R_{x;M}/k^*[x_1]},$$

$$\mathcal{D}_{x,y;M} = \mathcal{D}_{R_{x,y;M}/k^*[y_1]}.$$

Now, the divisor $\wp$ is trivial on k^*, hence is a prime divisor of K/k^*. Furthermore, we have $v_\wp(x_1) \geq 0$, $v_\wp(y_1) \geq 0$. Hence, by the case $r = 1$, and observing that $v_\wp(\mathcal{D}_{x;M}) = v_\wp(\mathcal{D}_x)$, $v_\wp(\mathcal{D}_{x,y;M}) = v_\wp(\mathcal{D}_{x,y})$, we find that

$$v_\wp(D^*_{x_1} y_1) = v_\wp(\mathcal{D}_{x,y}) - v_\wp(\mathcal{D}_x). \tag{2}$$

Similarly, let N denote the multiplicative set $k[y_1] - \{0\}$, let $R_{x,y;N}$ and $R_{y,N}$ be the quotient rings, with respect to N, of $R_{x,y}$ and R_y respectively, and let $\mathcal{D}_{x,y;N} = R_{x,y;N} \cdot \mathcal{D}_{x,y}$, $\mathcal{D}_{y,N} = R_{y,N} \cdot \mathcal{D}_y$. Then $R_{x,y;N}$ is the integral closure of $k_1[x_2,\ldots,x_r]$ in K, $R_{y;N}$ is the integral closure of $k_1[y_2,\ldots,y_r]$ in K, and we have

$$\mathcal{D}_{x,y;N} = \mathcal{D}_{R_{x,y;N}/k_1[x_2,\ldots,x_r]},$$

$$\mathcal{D}_{y;N} = \mathcal{D}_{R_{y;N}/k_1[y_2,\ldots,y_r]}.$$

Our prime divisor $\wp$ is also a prime divisor of K/k_1, and it is of the first kind with respect to the rings $k_1[x_2,\ldots,x_r]$,

$k_1[y_2,\ldots,y_r]$. Hence, by our induction hypothesis, and observing that $v_{\wp}(\vartheta_{x,y;N}) = v_{\wp}(\vartheta_{x,y})$, $v_{\wp}(\vartheta_{y,N}) = v_{\wp}(\vartheta_y)$, we find that

$$(3) \qquad v_{\wp}\left(\frac{\partial_1(y_2,\ldots,y_r)}{\partial_1(x_2,\ldots,x_r)}\right) = v_{\wp}(\vartheta_y) - v_{\wp}(\vartheta_{x,y}).$$

The proposition now follows from (1), (2), and (3).

<u>Prop. 15.2</u>: Let the notations and assumptions be the same as in Prop. 15.1. If k is algebraically closed, then $v_{\wp}(dy_1dy_2\ldots dy_r) = v_{\wp}(\vartheta_y)$.

Proof: We can choose $\{x_1,\ldots,x_r\}$ to be a set of uniformizing coordinates of $\wp$, and we can assume x_1 is a uniformizing parameter of $\wp$ (i.e., $v_{\wp}(x_1) = 1$). If $\bar{x}_2,\ldots,\bar{x}_r$ are the $\wp$-residues of $x_2,\ldots,x_r$, then the residue field Δ of $\wp$ is a separable algebraic extension of $k(\bar{x}_2,\ldots,\bar{x}_r)$. We shall identify x_i and $\bar{x}_i$ for $i = 2,\ldots,r$. Then, as before, we can view $\wp$ as a prime divisor of K/k^* where $k^* = k(x_2,\ldots,x_r)$. Since $v_{\wp}(x_1) = 1$ and since the residue field of $\wp$ is separable algebraic over k^*, the prime divisor $\wp$ of K/k^* is unramified over $k^*(x_1)$. Hence $\wp$ does not divide the different $\vartheta_{x,M}$, i.e., we have $v_{\wp}(\vartheta_x) = 0$. Since the x_i are uniformizing coordinates, we have $v_{\wp}(dy_1\ldots dy_r) = v_{\wp}\left(\frac{\partial(y_1,\ldots,y_r)}{\partial(x_1,\ldots,x_r)}\right) = v_{\wp}(\vartheta_y) - v_{\wp}(\vartheta_x)$ where the last equality follows from Prop. 14.10. Hence $v_{\wp}(dy_1\ldots dy_r) = v_{\wp}(\vartheta_y)$.

<u>Theorem 15.3</u>: Let V_a be an irreducible, r-dimensional affine hypersurface defined over an algebraically closed ground field k. Let $f(X_1,\dots,X_{r+1}) = 0$ be the equation of V_a. Let $R_o = k[x_1,\dots,x_{r+1}]$, let R be the integral closure of R_o in k(V) and let $\mathcal{L}$ be the conductor of R in R_o. Assume that $\{x_1,\dots,x_r\}$ is a separating transcendence basis of k(V)/k, and let $\omega = A dx_1 \dots dx_r$ be an r-fold differential. Then ω is regular on V_a if, and only if, $Af'_{r+1} \in \mathcal{L}$ where $f_i' = \partial f/\partial x_i$.

Proof: Let $\wp$ stand for a prime divisor of k(V)/k, of the first kind with respect to V_a (equivalently: $\wp$ is of the first kind with respect to $k[x_1,\dots,x_r]$). The following assertions are clearly equivalent:

(a) ω is regular on V_a.

(b) $v_\wp(\omega) \geq 0$ for all $\wp$.

(c) $v_\wp(A\vartheta_x) \geq$ for all $\wp$.

(d) $v_\wp(A\vartheta_x\mathcal{L}) \geq v_\wp(\mathcal{L})$ for all $\wp$.

(e) $v_\wp(Af'_{r+1}) \geq v_\wp(\mathcal{L})$ for all $\wp$. [By reduction to the case $r = 1$ we see at once that $v_\wp(\vartheta_x\mathcal{L}) = v_\wp(f'_{r+1})$].

Since $\mathcal{L}$ is pure (r-1)-dimensional, we have $\mathcal{L} = \bigcap_\wp \wp^{(v_\wp(\mathcal{L}))}$ where $\wp$ runs through all the minimal prime ideals in $R_x = k[x_1,\dots,x_r]$ and where $\wp^{(v_\wp(\mathcal{L}))}$ is the symbolic prime power. Thus (e) is equivalent to

(f) $Af'_{r+1} \in \mathcal{L}$.

Theorem 15.3 shows that any r-fold differential which is regular on the affine variety V_a can be written in the form $(\phi/f'_{r+1}).dx_1dx_2...dx_r$, where $\phi \in \mathcal{L}$; and conversely.

Let V/k be a hypersurface in projective S_{r+1}, let V_a be an affine representative of V and let P be a point of V. Let $y_0,y_1,...,y_n$ be strictly homogeneous coördinates of the general point of V/k, let $x_1,x_2,...,x_n$ be the non-homogeneous coördinates of that general point (where we may assume that $x_i = y_i/y_0$) and let $\mathfrak{o}$ be the local ring of P on V/k.

An element w of the function field k(V) is said to be a subadjoint function, locally at P, if w belongs to the conductor, in $\mathfrak{o}$, of the integral closure of $\mathfrak{o}$ in k(V). It follows that a function which is subadjoint, locally at P, belongs to the loca ring of P.

An element w of k(V) is said to be a subadjoint function of the affine representative V_a if w is locally subadjoint at every point of V_a. Since the intersection of the local rings of all th points of V_a is the coördinate ring $k[x_1,x_2,...,x_n]$ of V_a/k, a subadjoint function w of V_a/k can be written as a polynomial expression of $x_1,x_2,...,x_n$, with coefficients in k. A polynomia $\psi(X_1,X_2,...,X_n)$ in n indeterminates X_i, with coefficients in k, will be said to be a subadjoint polynomial of V_a/k if $\psi(x_1,x_2,...,x_n)$ is a subadjoint function of V_a/k. It is clear that an element w of $k[x_1,x_2,...,x_n]$ is a subadjoint function of V_a/k if and only if w belongs to the conductor $\mathcal{L}(V_a)$.

Finally, a homogeneous element w^* of the coördinate ring $k[y_0,y_1,\dots,y_n]$ will be said to be a subadjoint (homogeneous) function of the (projective) hypersurface V if w^* belongs to the conductor $\mathcal{L}(V)$; and a homogeneous polynomial $\phi(Y_0,Y_1,\dots,Y_{n+1})$ in the n+1 indeterminates Y_i, with coefficients in k, will be said to be a subadjoint form of V/k if $\phi(y_0,y_1,\dots,y_n)$ is a subadjoint function of V/k.

We consider an irredundant decomposition of $\mathcal{L}(V)$ into (homogeneous) primary components and we denote by $\mathcal{L}_a^*$ the intersection of those primary components whose prime ideals do not contain y_0. It is then immediate that if $\psi(X_1,X_2,\dots,X_n)$ is a polynomial of degree ν, with coefficients in k, then $\psi(x_1,x_2,\dots,x_n)$ is a subadjoint function of V_a/k if and only if the homogeneous function $\phi(y_0,y_1,\dots,y_n) = y_0^{\nu}\psi(y_1/y_0, y_2/y_0,\dots, y_n/y_0)$ belongs to $\mathcal{L}_a^*$. The function $\phi(y_0,y_1,\dots,y_n)$ is not necessarily itself a subajoint function of the (projective) hypersurface, but since the primary components of $\mathcal{L}(V)$ which are not included in $\mathcal{L}_a^*$ all contain y_0, it is clear that $y_0^{\mu}\phi(y_0,y_1,\dots,y_n) \in \mathcal{L}(V)$ for all sufficiently large integers μ. It follows that every subadjoint polynomial $\psi(X_1,X_2,\dots,X_n)$ of the affine variety V_a/k comes from a subadjoint form $\phi(Y_0,Y_1,\dots,Y_n)$ of V/k by setting $Y_0 = 1$ and $Y_i = X_i$ for $i > 0$; but it may be necessary to take for ϕ a subadjoint form of degree $\nu+\mu$ greater than the degree ν of ψ (and in that case Y_0^{μ} will be a factor of $\phi(Y)$). Conversely, it is obvious that if

$\phi(Y_o,Y_1,\ldots,Y_n)$ is a subadjoint form of V/k, of degree g, then $\phi(1,X_1,X_2,\ldots,X_n)$ is a subajoint polynomial of V_a/k (of degree $\leqq g$).

Let $F(Y_o,Y_1,\ldots,Y_{r+1}) = 0$ be the irreducible homogeneous equation of V/k, where $F(Y)$ is a form of degree n, and let $f(X_1,X_2,\ldots,X_{r+1}) = F(1,X_1,X_2,\ldots,X_{r+1})$ so that $f(X) = 0$ is the irreducible equation of V_a/k. We assume that k is algebraically closed and that $\{ x_1,x_2,\ldots,x_r \}$ is a separating transcendenee basis of $k(V)/k$.

Theorem 15.4. A necessary and sufficient condition that a differential

$$\omega = (A/f'_{r+1}(x))dx_1dx_2\ldots dx_r \ , \qquad A \in k(V),$$

be regular on the (projective) hypersurface V/k is that A be of the form $\psi(x)$, where $\psi(x)$ is a polynomial of degree $\leqq$ n-2-r such that $Y_o^{n-2-r}\ \psi(Y_1/Y_o,\ Y_2/Y_o,\ldots,Y_{r+1}/Y_o)$ is a subadjoint form of V/k.

Proof: We note the relations

$$f'_i\ dx_1dx_2\ldots dx_r = (-1)^{r+i-1}\ dx_1dx_2\ldots\widehat{dx_i}\ldots dx_{r+1}\ ,$$

$$i = 1,2,\ldots,r,$$

where the sign $\wedge$ above dx_i signifies that this differential factor dx_i has to be deleted. From these relations it follows at once that if we make a change of variables:

$$z_i = x_i + c_i x_{r+1}, \qquad i = 1,2,\ldots,r; \qquad c_i \in k;$$

$$z_{r+1} = x_{r+1},$$

and if $g(z_1, z_2, \ldots, z_{r+1}) = 0$ is the irreducible relation between the z_i, then

$$f'_{r+1}\, dz_1 dz_2 \ldots dz_r = g'_{r+1} \cdot dx_1 dx_2 \ldots dx_r .$$

Hence in the new expression of the differential ω the adjoint polynomial $\psi_1(z)$ which will occur is merely the transform of the polynomial $\psi(x)$. For a "non-special" value of the r constants c_i the term z_{r+1}^n will actually occur in $g(z)$. We thus may assume that the term X_{r+1}^n occurs in the original equation $f(X) = 0$ of V_a. That means that the point $Y_o = Y_1 = \ldots = Y_r = 0$, $Y_{r+1} = 1$ does not belong to V. Hence, if we denote by V_i the affine representative $V - V \cap H_i$ of V, where H_i is the hyperplane $Y_i = 0$ $(i = 0,1,\ldots,r)$, then <u>the</u> r+1 <u>affine varieties</u> V_i <u>cover</u> V. Then ω is regular on V if and only if it is regular on each V_i. For ω to be regular on V_o it is necessary and sufficient that A be of the form $\psi(x)$, where $\psi(x)$ is an adjoint polynomial of V_o $(= V_a)$. Among the different adjoint polynomials $\psi(X)$ such that $A = \psi(x)$ we take one of smallest possible degree h, and we let $\phi(Y) = Y_o^h \psi(Y_1/Y_o, Y_2/Y_o, \ldots, Y_{r+1}/Y_o)$. We have to show that ω is regular on each of the r affine varieties $V_1, V_2, \ldots, V_r$ if and only if $h \leqq n-2-r$ and $Y_o^{n-2-r-h}\, \phi(Y)$ is a subadjoint form of V/k.

We assume then that ω is regular on V and we shall show that the assumption that $k > n-2-r$ leads to a contradiction. Let, then, $\rho = h-(n-2-r) > 0$.

We set $\bar{x}_1 = 1/x_1 = y_0/y_1$, $\bar{x}_i = x_i/x_1 = y_i/y_1$, for $i = .,2,3,\ldots,r+1$. Then $(\bar{x}_1,\bar{x}_2,\ldots,\bar{x}_{r+1})$ is a general point of V_1/k, and the irreducible equation of V_1/k is $g(\bar{X}_1,\bar{X}_2,\ldots,\bar{X}_{r+1}) = 0$, where $g(X) = F(\bar{X}_1,1,\bar{X}_2,\ldots,\bar{X}_{r+1})$. Furthermore, $\{\bar{x}_1,\bar{x}_2,\ldots,\bar{x}_r\}$ is obviously a separating transcendence basis of $k(V)/k$. A simple calculation shows that

$$(1)\qquad \omega = \frac{\bar{x}_1^{\,n-r-2-h}\,\bar{\psi}(\bar{x})}{g'_{r+1}} \cdot d\bar{x}_1 d\bar{x}_2 \ldots d\bar{x}_r ,$$

where $\bar{\psi}(\bar{x}) = \phi(y)/y_1 h = \psi(x)\bar{x}_1^{\,h}$. We have therefore

$\omega = \dfrac{\bar{\psi}(\bar{x})}{\bar{x}_1^{\,\rho}\, g'_{r+1}}\, d\bar{x}_1 d\bar{x}_2 \ldots d\bar{x}_r$, $\rho > 0$. Since ω is regular on V_1 it follows, by Theorem 15.3, that there must exist a form $A_s(y)$, of degree $s \geqq 0$, such that $\bar{\psi}(\bar{x})/\bar{x}_1^{\,\rho} = A_s(y)/y_1^{\,s}$. Hence $\phi(y)y_1^{\,s}/y_1^{\,h} = A_s(y)\, y_0^{\,\rho}/y_1^{\,\rho}$. This implies that $y_1^{\,\nu}\phi(y)$ is divisible by y_0 in $k[y]$, for some integer $\nu \geqq 0$. If we repeat the same reasoning for the affine varieties $V_2,V_3,\ldots,V_r$, we find similarly that $y_i^{\,\nu}\phi(y)$ is divisible by y_0 in $k[y]$, for some integer $\nu \geqq 0$ and for $i = 0,1,2,\ldots,r$. By a theorem of Macaulay, the ideal $(F(Y),Y_0)$ in $k[Y]$ is unmixed, of projective dimension $r-1 \geqq 0$, and the same is true therefore also of the principal ideal (y_0) in $k[y]$. On the other hand, since the point $Y_0 = Y_1 = \ldots = Y_r = 0$, $Y_{r+1} = 1$, does not lie on V, the ideal

$(y_0, y_1, \ldots, y_r)$ is irrelevant (of projective dimension -1). It follows that $\phi(y)$ must be divisible by y_0, in contradiction with our choice of $\psi(X)$.

Having proved the inequality $h \leq n-2-r$, we can now write $\psi(x) = \phi(y)/y_0^{n-2-r}$, where we now drop the condition that the degree of $\psi(X)$ be minimal. Thus, now $\phi(Y)$ is a form of degree n-2-r. Then (1) shows that ω is regular on V if and only if $\phi(y)/y_i^{n-2-r}$ is a subadjoint function of the affine variety V_i, for $i = 0,1,\ldots,r$. Since these affine varieties cover V, it follows that ω is regular on V if and only if $\phi(y)$ is a subadjoint form of V/k. This completes the proof of the theorem.

<u>Cor. 15.5</u>: If V' is a normalization of V, then $p_g(V') = 1 + \dim_k \mathcal{L}_{n-r-2}(V)$, i.e., $p_g(V')$ is equal to the number of linearly independent subadjoint forms $\phi(Y)$ of V of degree n-r-2.

<u>Note 1</u>: If $n < r+2$, then $p_g(V') = 0$.

<u>Note 2</u>: The regularity of a differential ω, <u>on</u> V, does not, of course, imply that ω is a regular differential <u>of the field</u> k(V)/k, unless V satisfies some further conditions (as, for instance, the condition that the normalization of V be a non-singular variety). The special adjoint polynomials $\psi(X)$ such that the differential $\frac{\psi(x)}{f'_{r+1}} dx_1 dx_2 \ldots dx_r$ is regular for each valuation whose center lies on the affine variety V_a,

are called adjoint polynomials, and correspondingly one defines adjoint forms of V/k. The two concepts (of subadjoint and adjoint forms) coincide if the derived normal model of V/k is non-singular (in particular, if V has only so-called "ordinary singularities").

Let C/k be an irreducible curve in S_3 where k is algebraicall closed. We say that C is a complete intersection if I(C) in $k[Y_0,\dots,Y_3]$ has a basis of two elements. A similar definition holds in the affine case.

Theorem 15.6: Let C/k be an irreducible curve in affine A_3 which is the intersection of two surfaces $f(X,Y,Z) = 0$ and $g(X,Y,Z) = 0$. Let Q be a point of C and assume that one of the two surfaces has a simple point at Q. Let (x,y,z) be a general point of C/k, and let, say, x be a separating transcendental of k(C)/k. Then a necessary and sufficient condition for a differential ω of k(C)/k to be regular at Q is that be of the form

$$\omega = \frac{\phi}{\frac{\partial(f,g)}{\partial(y,z)}} dx$$

where $\phi \in \mathcal{L}_Q(C)$.

Proof: The following conditions can all be realized by means of a linear transformation and will therefore be assumed to be satisfied:

(1) Q is the origin.

(2) The line X = Y = 0 meets C only at Q.

(3) y is a primitive element of k(C)/k(x), and y is an integral function of x.

(4) z is integrally dependent on k[x,y].

(5) Z = 0 is the tangent plane of g(X,Y,Z) = 0 at Q where we are assuming Q is a simple point of g(X,Y,Z) = 0.

Let $\psi(x,y) = 0$ be the irreducible relation between x and y, and let D be the plane curve $\psi(X,Y) = 0$. Let P be the point (0,0) in the plane. Since $Q \in C$, we have $P \in D$. Let $\mathcal{O} = \mathcal{O}_Q(C/k)$, and let $\mathcal{O}_1 = \mathcal{O}_P(D/k)$. Clearly $\mathcal{O}_1 \subset \mathcal{O}$. We shall show $\mathcal{O}_1 = \mathcal{O}$.

Let v be a valuation of K = k(C) such that $\mathcal{O}_1 \subset R_v$, i.e., the center of v on D is P. Since $k[x,y] \subset \mathcal{O}_1$ and z is integral over k[x,y], we see that $k[x,y,z] \subset R_v$. Hence Q is the center of v on C because $v(x) > 0$ and $v(y) > 0$ and in view of condition (2). This shows that $\mathcal{O}$ is integrally dependent on $\mathcal{O}_1$.

Let $\mathfrak{m}$ be the maximal ideal of $\mathcal{O}$, i.e., $\mathfrak{m} = \mathcal{O}(x,y,z)$. We shall show that $\mathfrak{m} = \mathcal{O}(x,y)$. Consider the ideal (f,g,X,Y). This ideal has only one zero, namely Q. Therefore the ideal (f,g,X,Y) is primary for the ideal (X,Y,Z). We can write

$$g(X,Y,Z) = Zh(Z) + A(X,Y,Z) + B(X,Y,Z)Y$$

where $h(0) \neq 0$. Thus $(f,g,X,Y) = (f,Zh(Z),X,Y)$. Since $h(0) \neq 0$, we see that $h(Z) \notin (X,Y,Z)$. Hence $Z \in (f,Zh(Z),X,Y)$ and so $(f,g,X,Y) = (X,Y,Z)$. It follows that $(x,y) = (x,y,z)$ in $k[x,y,z]$.

Let $u \in \mathcal{O}$. For each $i \geq 1$ there exists a polynomial $p_i(x,y) \in k[x,y]$ such that $u - p_i(x,y) \in \mathfrak{m}^{i+1}$. Since $\mathcal{O}$ is integrally dependent on $\mathcal{O}_1$, $\mathcal{L}(\mathcal{O}/\mathcal{O}_1)$ contains a power of $\mathfrak{m}$. Hence, for large i, $\mathfrak{m}^{i+1} \subset \mathcal{O}_1$. Therefore $\mathcal{O} = \mathcal{O}_1$.

Since $\mathcal{O} = \mathcal{O}_1$, we have $\mathcal{L}_Q = \mathcal{L}_P$.

The statement "ω is regular at Q" is equivalent to each of the statements

(a) ω is regular at P.

(b) $\omega = Adx$ where $A\psi_y' \in \mathcal{L}_Q = \mathcal{L}_P$.

Since $\mathcal{O} = \mathcal{O}_1$, we can write $z = H_1(x,y)/H_0(x,y)$ where $H_0(0,0) \neq 0$. Let $h(X,Y,Z) = ZH_0(X,Y) - H_1(X,Y)$. We have

$$h(X,Y,Z) = A(X,Y,Z)f(X,Y,Z) + B(X,Y,Z)g(X,Y,Z).$$

Now either $A(0,0,0) \neq 0$ or $B(0,0,0) \neq 0$. Assume $A(0,0,0) \neq 0$. We have $\partial(h,g)/\partial(y,z) = A(x,y,z)$. $\partial(f,g)/\partial(y,z)$ and $A(x,y,z)$ is a unit in $\mathcal{O}_1$. It remains to prove that $\partial(h,g)/\partial(y,z)$ and ψ_y' are associates in $\mathcal{O}$.

Locally at Q we have $\psi(X,Y) = ah + bg$. Replace Z by H_1/H_0. Then locally at P we have $\psi(X,Y) \equiv 0 \pmod{g(X,Y, H_1/H_0)}$. Since $g(x,y,z) = 0$ we have $g(X,Y,H_1/H_0) \equiv 0 \pmod{\psi(X,Y)}$ locally at P.

Therefore $b(x,y,z)$ is a unit in $\mathcal{O}$. Differentiating $\psi(X,Y) = ah + bg$, we obtain

$$\psi'_y = a(x,y,z)h'_y + b(x,y,z)g'_y \quad \text{and}$$
$$0 = a(x,y,z)h'_z + b(x,y,z)g'_z.$$

Hence $\psi'_y = \frac{b(x,y,z)}{h'_z} \frac{\partial(h,g)}{\partial(y,z)}$. Now $h'_z = H_0(x,y)$ which is a unit in $\mathcal{O}$. Therefore $b(x,y,z)/h'_z$ is a unit in $\mathcal{O}$ and so ψ'_y and $\partial(h,g)/\partial(y,z)$ are associates in $\mathcal{O}$.

§16. Proof of the fundamental lemma.

Let V be a normal variety. We shall determine a certain divisorial cycle Δ such that, for all $q \geq 1$, $L_q - \Delta$ is complete. (Since V is normal, we know L_q is complete for large q.) Let dim V = r, and let $(y_0,\ldots,y_m)$ be a general point of V. By the Noether normalization lemma we know there exist elements $z_0,\ldots,z_{r+1}$ where $z_i = \sum_{j=0}^{m} c_{ij}y_j$, $c_{ij} \in k$, with the properties

(1) $k(y) = k(z)$.

(2) $k[y] = k[V]$ is integral over $k[z]$.

Now $k[z]$ may be regarded as the coordinate ring of a hypersurface V^* in S_{r+1} where V^* is the projection of V into S_{r+1} from a suitable S_{m-r-2}.

Let $R = k[y] = k[V]$, let $R^* = k[z] = k[V^*]$, and let $\bar{R}$ be the integral closure of R (hence also of R^*). Let $\mathcal{L}$ be the conductor of $\bar{R}$ in R^*, i.e., $\mathcal{L} = \mathcal{L}(V^*)$. Then $\mathcal{L} = \mathfrak{P}_1^{(\nu_1)} \cap \ldots \cap \mathfrak{P}_h^{(\nu_h)}$

where the $\bar{\mathfrak{p}}_i$ are minimal homogeneous prime ideals in $\bar{R}$. Since V is normal, each $\bar{\mathfrak{p}}_i$ determines a unique prime divisorial cycle Γ_i on V. Let $\Delta = \nu_1 \Gamma_1 + \ldots + \nu_h \Gamma_h$.

Prop. 16.1: If n is the order of V, then

$$L_q - \Delta = \bar{L}_q - \Delta = \{Z_\phi - \Delta \mid \phi \in \mathcal{L}_q\} = |K+(q-n+r+2)C_1|.$$

In particular, for $q = n-r-2$, $L_q - \Delta$ is the canonical system on V.

Proof: If $\phi \in \bar{R}_q$, then $\Delta \leq Z_\phi$ if, and only if, $\phi \in \mathcal{L}_q$. If $\phi \in \mathcal{L}_q$, then $Z_\phi \in L_q$ since $\mathcal{L}_q \subseteq R_q^* \subseteq R_q$. Thus $L_q - \Delta$ $\bar{L}_q - \Delta = \{Z_\phi - \Delta \mid \phi \in \mathcal{L}_q\}$; and since $\bar{L}_q$ is complete, this means $L_q - \Delta$ is complete for $q \geq 1$.

Since n is the order of V, it is also the order of V^*. Hence any regular differential ω on V^* is of the form

$$\omega = \frac{\phi(z)}{z_0^{n-r-2}\, f'_{r+1}}\, d(\frac{z_1}{z_0}) \ldots d(\frac{z_r}{z_0})$$

where $\phi \in \mathcal{L}_{n-r-2}$. Thus "$\omega$ regular on V" implies that the divisor of ω, (ω), is in $L_q - \Delta = \{Z_\phi - \Delta\}$ since for any prime divisorial cycle Γ on V for which $z_0 \neq 0$ we have

$$\begin{aligned} v_\Gamma(\omega) &= v_\Gamma(\phi(z)/z_0^{n-r-2}) + v_\Gamma(d(z_1/z_0)\ldots d(z_r/z_0)) - v_\Gamma(f'_{r+1}) \\ &= v_\Gamma(Z_\phi) + v_\Gamma(\vartheta_z) - v_\Gamma(f'_{r+1}) \\ &= v_\Gamma(Z_\phi) - v_\Gamma(\mathcal{L}) \\ &= v_\Gamma(Z_\phi) - v_\Gamma(\Delta). \end{aligned}$$

If $z_0 = 0$ on Γ, we can find some z_1 such that $z_1 \neq 0$ on Γ and repeat the argument above again obtaining $v_\Gamma(\omega) = v_\Gamma(Z_\phi) - v_\Gamma(\Delta$

Thus $|K| \subset L_{n-r-2} - \Delta$ and hence $|K| = L_{n-r-2} - \Delta$, since both systems are complete. The assertion $L_q - \Delta = |K + (q-n+r+2)C_1|$ is an obvious consequence.

Let F be a non-singular surface in S_n with general point $(y_o,\ldots,y_n)$. We project F onto a surface F^* in S_3 where we can assume that:

(1) (y_o,y_1,y_2,y_3) is a general point of F^*.

(2) F and F^* are birationally equivalent.

(3) $k[F]$ is integral over $k[F^*]$.

Let $f^*(Y_o,Y_1,Y_2,Y_3) = 0$ be the equation of F^*.

For each integer $m \geq 1$, we can choose hypersurfaces G_m in S_n defined by a form $g_m^*(Y_o,Y_1,Y_2,Y_3)$ of degree m so that the following conditions are satisfied.

(a) The surface G_m^* in S_3 defined by $g_m^*(Y) = 0$ is non-singular.

(b) $F \cap G_m$ is an irreducible non-singular curve C_m, and moreover C_m is the complete intersection of F and G_m, i.e., the principal ideal $(g_m^*(y))$ in $k[F]$ differs from the ideal of C_m in $k[F]$ at most by an irrelevant primary component.

(c) If $(\bar{y}_o,\ldots,\bar{y}_n)$ is the general point of C_m, and C_m^* is the locus of $(\bar{y}_o,\bar{y}_1,\bar{y}_2,\bar{y}_3)$ in S_3, then

(2') C_m and C_m^* are birationally equivalent.

(3') $k[C_m]$ is integral over $k[C_m^*]$.

(d) C_m^* is a complete intersection of F^* and the surface G_m^*.

Actually it can be shown that (d) is implied by (a), (b), and (c).

Prop. 16.2: If $\phi(y_0,y_1,y_2,y_3) \in \mathcal{L}(F^*)$, then $\phi(\bar{y}_0,\bar{y}_1,\bar{y}_2,\bar{y}_3) \in \mathcal{L}(C_m^*$

Proof: We first make the following remark: Let C_m be an irreducible curve on a non-singular surface F. In general, if $Q \in C_m$ and if Q is simple for both C_m and F, then we can choose t and τ as uniformizing parameters of Q on F such that t = 0 is a local equation of C_m at Q. Let ω be a differential of degree two on F and assume ω is regular at Q. Then $\mathrm{Tr}_{C_m}^t \omega = \bar{\omega}$ is a regular differential of degree one at Q because $\omega = A\,d\tau\,dt$ and $\bar{\omega} = \bar{A}\,d\bar{\tau}$ where $A \in \mathcal{O}_Q(F/k)$ and $\bar{A} \in \mathcal{O}_Q(C_m/k)$. It will be sufficient to consider the various affine representatives determined by $Y_i = 0$, $i = 0,1,2,3$.

Let $F_a^* = F^* - (F^* \cap H_o)$ and $C_{m,a}^* = C_m^* - (C_m^* \cap H_o)$ where H_o is the hyperplane $Y_o = 0$. Let $x_i = y_i/y_o$ and $\bar{x}_i = \bar{y}_i/\bar{y}_o$, $i = 1,2,3$. Th equation of F_a^* is $f(X_1,X_2,X_3) = f^*(1,X_1,X_2,X_3) = 0$, and $C_{m,a}^*$ is a complet intersection of $f(X) = 0$ and $g_m(X) = 0$ where $g_m(X) = g_m^*(1,X)$.

Without loss of generality we can assume $\{x_1,x_2\}$ and $\{\bar{x}_1\}$ are separating transcendence bases of $k(x_1,x_2,x_3)/k$ and $k(\bar{x}_1,\bar{x}_2,\bar{x}_3)/$ respectively.

Let $\psi(x) = \phi(1,x_1,x_2,x_3)$. Then $\psi(x) \in \mathcal{L}(F_a^*)$ and if $\omega = (\psi(x)/f'_{x_3})dx_1dx_2$, then ω is regular on F_a^*. Let $t = g_m(x)$, and let F_a be the affine representative of F with respect to $Y_o =$ (If Q is any point of $C_m \cap F_a$, then t = 0 is a local equation of C_m at Q. Let $\bar{\omega} = \mathrm{Tr}_{C_m}^t \omega$. Then $\bar{\omega}$ is regular on $C_{m,a}^*$. It is easy to see that t and x_1 are uniformizing coordinates of C_m. Hence we can write $\omega = A\,dx_1 dt$ and $\bar{\omega} = \bar{A}\,d\bar{x}_1$. We have $dx_1 dt = (\frac{\partial t}{\partial x_2}) dx_1 dx_2$, and $\frac{\partial t}{\partial x_2} = g'_{x_2} + g'_{x_3}(\frac{\partial x_3}{\partial x_2})$. This yields

$$dx_1 dt = \frac{-\begin{vmatrix} f'_{x_2} & f'_{x_3} \\ g'_{x_2} & g'_{x_3} \end{vmatrix}}{f'_{x_3}} dx_1 dx_2 = -\frac{\frac{\partial(f,g)}{\partial(x_2,x_3)}}{f'_{x_3}} dx_1 dx_2.$$

Hence $= -\frac{\psi(x)}{\frac{\partial(f,g)}{\partial(x_2,x_3)}} dx_1 dt$ and therefore $\bar{\omega} = -\frac{\psi(\bar{x})}{\frac{\partial(f,g)}{\partial(\bar{x}_2,\bar{x}_3)}} d\bar{x}_1$.

Since $\bar{\omega}$ is regular on $C^*_{m,a}$, we have $\psi(\bar{x}) \in \mathcal{L}(C^*_{m,a})$ which proves the proposition.

Let Δ be the divisorial cycle on F defined by $\mathcal{L}(F^*)(\Delta \geq 0)$, and let $\bar{\Delta}_m$ be the divisorial cycle on C_m defined by $\mathcal{L}(C^*_m)$. We know $Tr_{C_m} \mathcal{L}(F^*)$ is an ideal contained in $\mathcal{L}(C^*_m)$ where we note that $Tr_{C_m} \mathcal{L}(F^*) \neq (0)$. The divisor on C_m defined by the homogeneous ideal $Tr_{C_m} \mathcal{L}(F^*)$ is the intersection cycle $\Delta . C_m$. Therefore $\Delta . C_m \geq \bar{\Delta}_m$.

Let L_q be the system of sections of F with hypersurfaces of order q and let $\bar{L}_{q,m}$ be the corresponding system on C_m. For all $q \geq 1$, $L_q - \Delta$ and $\bar{L}_{q,m} - \bar{\Delta}_m$ are complete. Therefore $\bar{L}_{q,m} - \Delta . C_m$ is complete for all $q \geq 1$. We must show that $Tr_{C_m}(L_q - \Delta)$ is complete for large m.

By Lemma 14.8 we know that for each q there exists an integer $N(q)$ such that if $D \in |L_q|$ and $D.C_m \geq \Delta .C_m$ where $D \geq 0$ and $m \geq N(q)$, then $D \geq \Delta$. Let $\overline{W} \in |Tr_{C_m}(L_q-\Delta)| = \overline{L}_{q,m} - \Delta .C_m$. Then $\overline{W} + \Delta .C_m$ is of the form $D.C_m$ where $D \in L_q$ and so $D.C_m \geq \Delta .C_m$ because $\overline{W} \geq 0$. Then if $m \geq N(q)$, $D = \Delta + E$ where $E \geq 0$. Hence $E.C_m = \overline{W}$ and $Tr_{C_m}(L_q-\Delta)$ is complete. This completes the proof of the F.L.

Offsetdruck: Julius Beltz, Weinheim/Bergstr

Lecture Notes in Mathematics

Bisher erschienen/Already published

Vol. 1: J. Wermer, Seminar über Funktionen-Algebren. IV, 30 Seiten. 1964. DM 3,80 / 0.95

Vol. 2: A. Borel, Cohomologie des espaces localement compacts d'après J. Leray. IV, 93 pages. 1964. DM 9,– / $ 2.25

Vol. 3: J. F. Adams, Stable Homotopy Theory. 2nd. revised edition. IV, 78 pages. 1966. DM 7,80 / $ 1.95

Vol. 4: M. Arkowitz and C. R. Curjel, Groups of Homotopy Classes. 2nd. revised edition. IV, 36 pages. 1967. DM 4,80 / $ 1.20

Vol. 5: J.-P. Serre, Cohomologie Galoisienne. Troisième édition. VIII, 214 pages. 1965. DM 18,– / $ 4.50

Vol. 6: H. Hermes, Eine Termlogik mit Auswahloperator. IV, 42 Seiten. 1965. DM 5,80 / $ 1.45

Vol. 7: Ph. Tondeur, Introduction to Lie Groups and Transformation Groups. VIII, 176 pages. 1965. DM 13,50 / $ 3.40

Vol. 8: G. Fichera, Linear Elliptic Differential Systems and Eigenvalue Problems. IV, 176 pages. 1965. DM 13,50 / $ 3.40

Vol. 9: P. L. Ivănescu, Pseudo-Boolean Programming and Applications. IV, 50 pages. 1965. DM 4,80 / $ 1.20

Vol. 10: H. Lüneburg, Die Suzukigruppen und ihre Geometrien. VI, 111 Seiten. 1965. DM 8,– / $ 2.00

Vol. 11: J.-P. Serre, Algèbre Locale. Multiplicités. Rédigé par P. Gabriel. Seconde édition. VIII, 192 pages. 1965. DM 12,– / $ 3.00

Vol. 12: A. Dold, Halbexakte Homotopiefunktoren. II, 157 Seiten. 1966. DM 12,– / $ 3.00

Vol. 13: E. Thomas, Seminar on Fiber Spaces. IV, 45 pages. 1966. DM 4,80 / $ 1.20

Vol. 14: H. Werner, Vorlesung über Approximationstheorie. IV, 184 Seiten und 12 Seiten Anhang. 1966. DM 14,– / $ 3.50

Vol. 15: F. Oort, Commutative Group Schemes. VI, 133 pages. 1966. DM 9,80 / $ 2.45

Vol. 16: J. Pfanzagl and W. Pierlo, Compact Systems of Sets. IV, 48 pages. 1966. DM 5,80 / $ 1.45

Vol. 17: C. Müller, Spherical Harmonics. IV, 46 pages. 1966. DM 5,– / $ 1.25

Vol 18: H.-B. Brinkmann und D. Puppe, Kategorien und Funktoren. XII, 107 Seiten, 1966. DM 8,– / $ 2.00

Vol. 19: G. Stolzenberg, Volumes, Limits and Extensions of Analytic Varieties. IV, 45 pages. 1966. DM 5,40 / $ 1.35

Vol. 20: R. Hartshorne, Residues and Duality. VIII, 423 pages. 1966. DM 20,– / $ 5.00

Vol. 21: Seminar on Complex Multiplication. By A. Borel, S. Chowla, C. S. Herz, K. Iwasawa, J.-P. Serre. IV, 102 pages. 1966. DM 8,– / $ 2.00

Vol. 22: H. Bauer, Harmonische Räume und ihre Potentialtheorie. IV, 175 Seiten. 1966. DM 14,– / $ 3.50

Vol. 23: P. L. Ivănescu and S. Rudeanu, Pseudo-Boolean Methods for Bivalent Programming. 120 pages. 1966. DM 10,– / $ 2.50

Vol. 24: J. Lambek, Completions of Categories. IV, 69 pages. 1966. DM 6,80 / $ 1.70

Vol. 25: R. Narasimhan, Introduction to the Theory of Analytic Spaces. IV, 143 pages. 1966. DM 10,– / $ 2.50

Vol. 26: P.-A. Meyer, Processus de Markov. IV, 190 pages. 1967. DM 15,– / $ 3.75

Vol. 27: H. P. Künzi und S. T. Tan, Lineare Optimierung großer Systeme. VI, 121 Seiten. 1966. DM 12,– / $ 3.00

Vol. 28: P. E. Conner and E. E. Floyd, The Relation of Cobordism to K-Theories. VIII, 112 pages. 1966. DM 9,80 / $ 2.45

Vol. 29: K. Chandrasekharan, Einführung in die Analytische Zahlentheorie. VI, 199 Seiten. 1966. DM 16,80 / $ 4.20

Vol. 30: A. Frölicher and W. Bucher, Calculus in Vector Spaces without Norm. X, 146 pages. 1966. DM 12,– / $ 3.00

Vol. 31: Symposium on Probability Methods in Analysis. Chairman. D. A. Kappos. IV, 329 pages. 1967. DM 20,– / $ 5.00

Vol. 32: M. André, Méthode Simpliciale en Algèbre Homologique et Algèbre Commutative. IV, 122 pages. 1967. DM 12,– / $ 3.00

Vol. 33: G. I. Targonski, Seminar on Functional Operators and Equations. IV, 110 pages. 1967. DM 10,– / $ 2.50

Vol. 34: G. E. Bredon, Equivariant Cohomology Theories. VI, 64 pages. 1967. DM 6,80 / $ 1.70

Vol. 35: N. P. Bhatia and G. P. Szegö, Dynamical Systems. Stability Theory and Applications. VI, 416 pages. 1967. DM 24,– / $ 6.00

Vol. 36: A. Borel, Topics in the Homology Theory of Fibre Bundles. VI, 95 pages. 1967. DM 9,– / $ 2.25

Vol. 37: R. B. Jensen, Modelle der Mengenlehre. X, 176 Seiten. 1967. DM 14,– / $ 3.50

Vol. 38: R. Berger, R. Kiehl, E. Kunz und H.-J. Nastold, Differentialrechnung in der analytischen Geometrie IV, 134 Seiten. 1967. DM 12,– / $ 3.00

Vol. 39: Séminaire de Probabilités I. II, 189 pages. 1967. DM 14,– / $ 3.50

Vol. 40: J. Tits, Tabellen zu den einfachen Lie Gruppen und ihren Darstellungen. VI, 53 Seiten. 1967. DM 6.80 / $ 1.70

Vol. 41: A. Grothendieck, Local Cohomology. VI, 106 pages. 1967. DM 10.– / $ 2.50

Vol. 42: J. F. Berglund and K. H. Hofmann, Compact Semitopological Semigroups and Weakly Almost Periodic Functions. VI, 160 pages. 1967. DM 12,– / $ 3.00

Vol. 43: D. G. Quillen, Homotopical Algebra VI, 157 pages. 1967. DM 14,– / $ 3.50

Vol. 44: K. Urbanik, Lectures on Prediction Theory IV, 50 pages. 1967. DM 5,80 / $ 1.45

Vol. 45: A. Wilansky, Topics in Functional Analysis VI, 102 pages. 1967. DM 9,60 / $ 2.40

Vol. 46: P. E. Conner, Seminar on Periodic Maps IV, 116 pages. 1967. DM 10,60 / $ 2.65

Vol. 47: Reports of the Midwest Category Seminar I. IV, 181 pages. 1967. DM 14,80 / $ 3.70

Vol. 48: G. de Rham, S. Maumary et M. A. Kervaire, Torsion et Type Simple d'Homotopie. IV, 101 pages. 1967. DM 9,60 / $ 2.40

Vol. 49: C. Faith, Lectures on Injective Modules and Quotient Rings. XVI, 140 pages. 1967. DM 12,80 / $ 3.20

Vol. 50: L. Zalcman, Analytic Capacity and Rational Approximation, VI, 155 pages. 1968. DM 13.20 / $ 3.40

Vol. 51: Séminaire de Probabilités II. IV, 199 pages. 1968. DM 14,– / $ 3.50

Vol. 52. D. J. Simms, Lie Groups and Quantum Mechanics. IV, 90 pages. 1968. DM 8,– / $ 2.00

Vol. 53: J. Cerf, Sur les difféomorphismes de la sphère de dimension trois $(\Gamma_4 = 0)$. XII, 133 pages. 1968. DM 12,– / $ 3.00

Vol. 54: G. Shimura, Automorphic Functions and Number Theory. VI, 69 pages. 1968. DM 8,– / $ 2.00

Vol. 55: D. Gromoll, W. Klingenberg und W. Meyer Riemannsche Geometrie im Großen VI, 287 Seiten. 1968. DM 20,– / $ 5.00

Bitte wenden / Continued

Vol. 56: K. Floret und J. Wloka,
Einführung in die Theorie der lokalkonvexen Räume
VIII, 194 Seiten. 1968. DM 16,– / $ 4.00

Vol. 57: F. Hirzebruch und K. H. Mayer,
O(n)-Mannigfaltigkeiten, exotische Sphären und Singularitäten.
IV, 132 Seiten. 1968. DM 10,80 / $ 2.70

Vol. 58: Kuramochi Boundaries of Riemann Surfaces.
IV, 102 pages. 1968. DM 9,60 / $ 2.40

Vol. 59: K. Jänich, Differenzierbare G-Mannigfaltigkeiten.
VI, 89 Seiten. 1968. DM 8,– / $ 2.00

Vol. 60: Seminar on Differential Equations and Dynamical Systems. Edited by G. S. Jones
VI, 106 pages. 1968. DM 9,60 / $ 2.40

Vol. 61: Reports of the Midwest Category Seminar II.
IV, 91 pages. 1968. DM 9,60 / $ 2.40

Vol. 62: Harish-Chandra, Automorphic Forms on Semisimple Lie Groups
X, 138 pages. 1968. DM 14,– / $ 3.50

Vol. 63: F. Albrecht, Topics in Control Theory.
IV, 65 pages. 1968. DM 6,80 / $ 1.70

Vol. 64: H. Berens, Interpolationsmethoden zur Behandlung von Approximationsprozessen auf Banachräumen.
VI, 90 Seiten. 1968. DM 8,– / $ 2.00

Vol. 65: D. Kölzow, Differentiation von Maßen.
XII, 102 Seiten. 1968. DM 8,– / $ 2.00

Vol. 66: D. Ferus, Totale Absolutkrümmung in Differentialgeometrie und -topologie. VI, 85 Seiten. 1968. DM 8,– / $ 2.00

Vol. 67: F. Kamber and P. Tondeur, Flat Manifolds.
IV, 53 pages. 1968. DM 5,80 / $ 1.45

Vol. 68: N. Boboc et P. Mustată, Espaces harmoniques associés aux opérateurs différentiels linéaires du second ordre de type elliptique.
VI, 95 pages. 1968. DM 8,60 / $ 2.15

Vol. 69: Seminar über Potentialtheorie.
Herausgegeben von H. Bauer.
VI, 180 Seiten. 1968. DM 14,80 / $ 3.70

Vol. 70: Proceedings of the Summer School in Logic.
Edited by M. H. Löb.
IV, 331 pages. 1968. DM 20,– / $ 5.00

Vol. 71: Séminaire Pierre Lelong (Analyse), Année 1967-1968.
VI, 19 pages. 1968. DM 14,– / $ 3.50

Vol. 72: The Syntax and Semantics of Infinitary Languages.
Edited by J. Barwise.
IV, 268 pages. 1968. DM 18,– / $ 4.50

Vol. 73: P. E. Conner, Lectures on the Action of a Finite Group.
IV, 123 pages. 1968. DM 10,– / $ 2.50

Vol. 74: A. Fröhlich, Formal Groups.
IV, 140 pages. 1968. DM 12,– / $ 3.00

Vol. 75: G. Lumer, Algèbres de fonctions et espaces de Hardy. VI, 80 pages. 1968. DM 8 – / $ 2.00

Vol. 76: R. G. Swan, Algebraic K-Theory.
IV, 262 pages. 1968. DM 18,– / $ 4.50

Vol. 77: P.-A. Meyer, Processus de Markov: la frontière de Martin. IV, 123 pages. 1968. DM 10,– / $ 2.50

Vol. 78: H. Herrlich, Topologische Reflexionen und Coreflexionen.
XVI, 166 Seiten. 1968. DM 12,- / $ 3.00

Vol. 79: A. Grothendieck, Catégories Cofibrées Additives et Complexe Cotangent Relatif.
IV, 167 pages. 1968. DM 12,– / $ 3.00

Vol. 80: Seminar on Triples and Categorical Homology Theory. Edited by B. Eckmann
IV, 398 pages. 1969. DM 20,– / $ 5.00

Vol. 81: J.-P. Eckmann et M. Guenin, Méthodes Algébriques en Mécanique Statistique.
VI, 131 pages. 1969. DM 12,– / $ 3.00

Vol. 82: J. Wloka, Grundräume und verallgemeinerte Funktionen
VIII, 131 Seiten. 1969. DM 12,– / $ 3.00